PREMELTONS IN DNA

A Unifying Polymer Physics Concept to Understand DNA Physical Chemistry and Molecular Biology

Henry M. Sobell

For Cover Illustration:

Stereochemistry of actinomycin D binding to DNA.

INKS & BINDINGS

Inks and Bindings
888-290-5218
www.inksandbindings.com
orders@inksandbindings.com

Preface

As a younger scientist, I could have never imagined that crystallographic studies of intercalation complexes would ever lead to understanding the mechanism of how drugs and dyes intercalate into DNA. Nor that these insights would ultimately come from concepts implicit in the use of nonlinear-mathematics to describe other areas of science, notably in both physics and engineering.

However, this is exactly what has happened, and more.

Thus, for example, the Bloch domain walls that separate spin-up spin-down domains in unmagnetized iron can be modeled through the use of the sine-Gordon equation, a nonlinear partial differential equation capable of exact solution. These solutions (known as the kink and antikink solutions) describe the existence of nonlinear phase-boundaries, each containing one hundred or more lattice-sites, which connect spin-up spin-down domains in the iron lattice. In the presence of an external magnetic field, these boundaries move frictionlessly throughout the lattice, converting spin-up to spin-down orientations (or, vice versa, depending on the direction of the external magnetic field), giving rise to ferromagnetism.

Another example is the origin of crystal-lattice dislocations (either of the edge or spiral type) which move frictionlessly in crystals at normal thermal energies. Again, these can be modeled through the use of the sine-Gordon equation to give the kink and the antikink solutions, which in this case, describe the presence of phase boundaries that exist before and after slippage. A particularly interesting application of this theory concerns the brittleness of iron. When subjected to either bending or shearing distortions, iron contains crystal-lattice dislocations that move and eventually coalesce, causing breakage. If carbon is added to molten iron, steel is formed. Steel is far more resilient to both bending and shearing distortions, this reflecting the ability of carbon (either as elemental carbon, or in chemical union with iron) to "pin" these crystal-lattice dislocations. This example exemplifies how the physical properties of metals are altered through the creation of alloys in metallurgical science.

It is with this background information in mind that I have written this book for my colleagues in molecular biology. Topics such as the physical nature of DNA- premelting and melting, structural phase-transitions interconverting A-, B- and Z-DNA, chain- slippage phenomena that give rise to single- and to double-stranded branch-migration, nuclease-hypersensitive sites present in both naked DNA and DNA in active and inactive chromatin, the structure of DNA within the promoter and terminator, and in extremely active genes – between transcription complexes, and finally, the way in which both actinomycin and echinomycin inhibit transcription – all can readily be understood through the concepts explained herein.

So - let us now examine the theory and the data that supports it.

HENRY M. SOBELL
Lake Luzerne, New York
January 2023

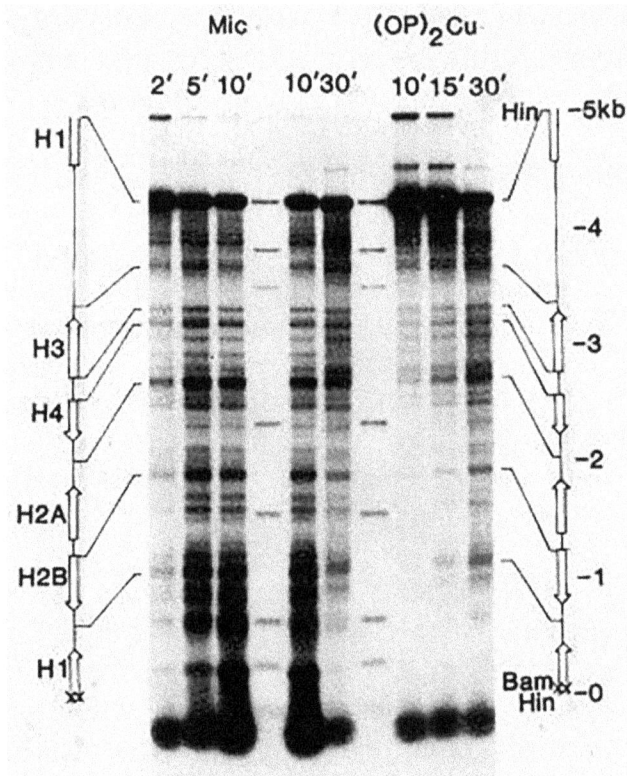

This study demonstrates the remarkable similarities between the chemical-nuclease, 1, 10- phenanthroline- copper (I), an intercalator, and the micrococcal-nuclease – in their ability to recognize and to cleave hypersensitive-sites in a 5,000 base-pair circular DNA fragment containing the histone gene-cluster from *D. melanogaster*. These nuclease hypersensitive-sites have been attributed by the author to reflect the presence of small premelted DNA regions (called, premeltons) that nucleate DNA melting at higher temperatures. This paper further explains the structural and physical nature of such regions, along with the implications they have in understanding DNA physical-chemistry and molecular-biology.

Abstract

Premeltons are examples of emergent-structures (i.e., structural-solitons) that arise spontaneously in DNA due to the presence of nonlinear-excitations in its structure. They are of two kinds.

B-B (or (A-A) premeltons form at specific DNA-regions to nucleate site-specific DNA melting. They are stationary and, being globally-nontopological, undergo breather-motions that allow drugs and dyes to intercalate into DNA.

B-A (or A-B) premeltons, on the other hand, are mobile, and being globally-topological, act as phase-boundaries transforming B- into A- DNA during the structural phase-transition. They are not expected to undergo breather-motions.

A key feature of both types of premeltons is the presence of an intermediate structural-form in their central regions (proposed as being a transition-state intermediate in DNA-melting and in the B- or A- transition), which differs from either A- or B- DNA. Called beta-DNA, this is both metastable and hyperflexible – and contains an alternating sugar-puckering pattern along the polymer-backbone combined with the partial-unstacking (in its lower- energy forms) of every other base-pair. Beta-DNA is connected to either B- or A- DNA on either side by boundaries possessing a gradation of nonlinear structural-change, these being called the kink and antikink regions.

The presence of premeltons in DNA leads to a unifying theory to understand much of DNA physical chemistry and molecular biology. In particular, premeltons are predicted to define the 5' and 3' ends of genes in naked-DNA and in DNA in active-chromatin, this having important implications for understanding physical-aspects of the initiation, elongation and termination of RNA-synthesis during transcription. For these and other reasons, the model will be of broader interest to the general audience working in these areas.

The model explains a wide variety of data and carries with it a number of experimental predictions – all readily testable – as described within.

Table of Contents

1. Introduction

The possibility that nonlinear excitations (solitons) exist in biopolymers and play a central role in energy-transfer was first advanced by Davydov in his classic-series of papers (1-3). A different class of solitons giving rise to localized conformational changes in DNA has also been advanced by Englander et al. (4), and subsequently by Krumhansl and Alexander (5), to explain DNA- breathing phenomena. A large number of papers have since appeared describing how the presence of nonlinear-excitations in DNA determine its early melting behavior -- notably within promoter regions (6-20).

Solitons are intrinsic, locally-coherent excitations that move along a polymer chain with a velocity significantly less than the speed of sound (they can even be stationary). They are combinations of intramolecular and deformational excitations, which appear as a consequence of an intrinsic nonlinear-instability in the polymer structure.

Extensive research on solitons in many physical systems has shown that this nonlinearity gives the spatially localized excitation a robust character (21-26). Solitons do not interact with conventional normal-mode excitations (i.e., phonons). They have their own identity and can be treated by Newtonian-dynamics as heavy Brownian-like particles, each having an "effective mass". Solitary structures – as sites for biochemical activity – behave like independent species and can be treated by statistical mechanics and chemical thermodynamics. They can arise from equilibrium or nonequilibrium processes.

The importance of the beta-DNA form (i.e., earlier proposed to be a key metastable and hyperflexible premelted DNA-structure that acts as a transition-state intermediate in both DNA- melting and in the B- to A- structural phase-transition), led the author and his research colleague, Asok Banerjee, to further propose the existence of collectively-localized nonlinear excitations (*solitons*) to exist in DNA-structure. These arise as a consequence of an intrinsic nonlinear-instability associated with interconversions between the two-predominant sugar- pucker conformations (i.e., C2' *endo* and C3' *endo*) accompanying base-pair unstacking. In their bound-state, soliton-antisoliton pairs surround small beta-DNA core regions, these regions capable of undergoing breather-motions which facilitate the intercalation of drugs and dyes into DNA. Additionally, such bound-state structures nucleate DNA-melting, and give rise to other kinds of structural phase-transitions in DNA [i.e., notably, both the B- to A- and the B- to Z- (or A- to Z-) structural phase-transitions]. They have been called, *premeltons* (27-32).

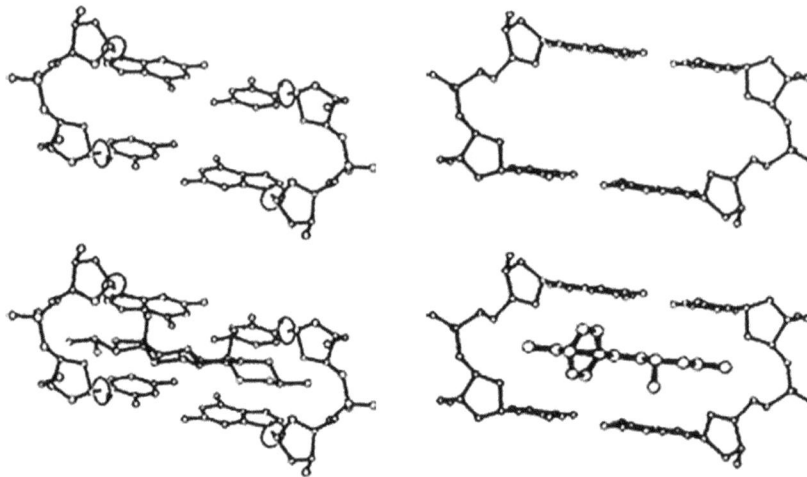

The beta-structural element in DNA alternating between its lowest- and highest-energy states within the centers of B-B (or A-A) premeltons, "pinned" and "unpinned" by irehdiamine and ethidium. This reflects the presence of dynamic breather-motions within premeltons, which facilitate the ability of drugs and dyes to intercalate into DNA.

More generally, premeltons (i.e., either of the B-B or A-A types) are proposed to arise spontaneously within the early melting-regions of DNA to nucleate site-specific melting. Their appearance explains the presence of micrococcal-nuclease and pancreatic DNase I hypersensitive-sites at the 5' and 3' ends of genes, both within naked DNA and in DNA in transcriptionally-active chromatin. Since central beta-structural elements within premeltons alternate between their lowest- and highest- energy states, they are able to act as substrates for both enzymes – pancreatic-DNase I cleaving its lowest-energy state, micrococcal-nuclease cleaving its highest-energy state. Additionally, both enzymes are capable of cleaving beta-structural elements statically present in both nucleosomal DNA and in linker-regions connecting nucleosomes in the higher-order solenoidal-structure of chromatin.

These concepts predict irehdiamine and ethidium to be competitive-inhibitors of the pancreatic-DNase I and micrococcal-nuclease cleavage-reactions, both in naked-DNA and in DNA in active- and inactive-chromatin. Experiments to test these and other predictions have been summarized in the concluding portion of this article.

2. Premeltons are (Low-Energy) Breather-Solitons in DNA

Conceptually, B-B (or A-A) premeltons are breather-solitons, these arising spontaneously within the early melting-regions of DNA to nucleate site-specific melting. They are stationary and being globally nontopological undergo breather-motions that allow drugs and dyes to intercalate into DNA. Premeltons have their own identity – they can be considered to be independent molecular-species, each consisting of a central metastable and hyperflexible beta- DNA core-region modulated into B- (or A-) DNA on either side through kink and antikink boundaries[1]. These boundaries act as energy domain-walls, capable of moving in and out with minimal energy-dissipation [2, 3].

See Figure 1.

PREMELTON BREATHER DYNAMICS

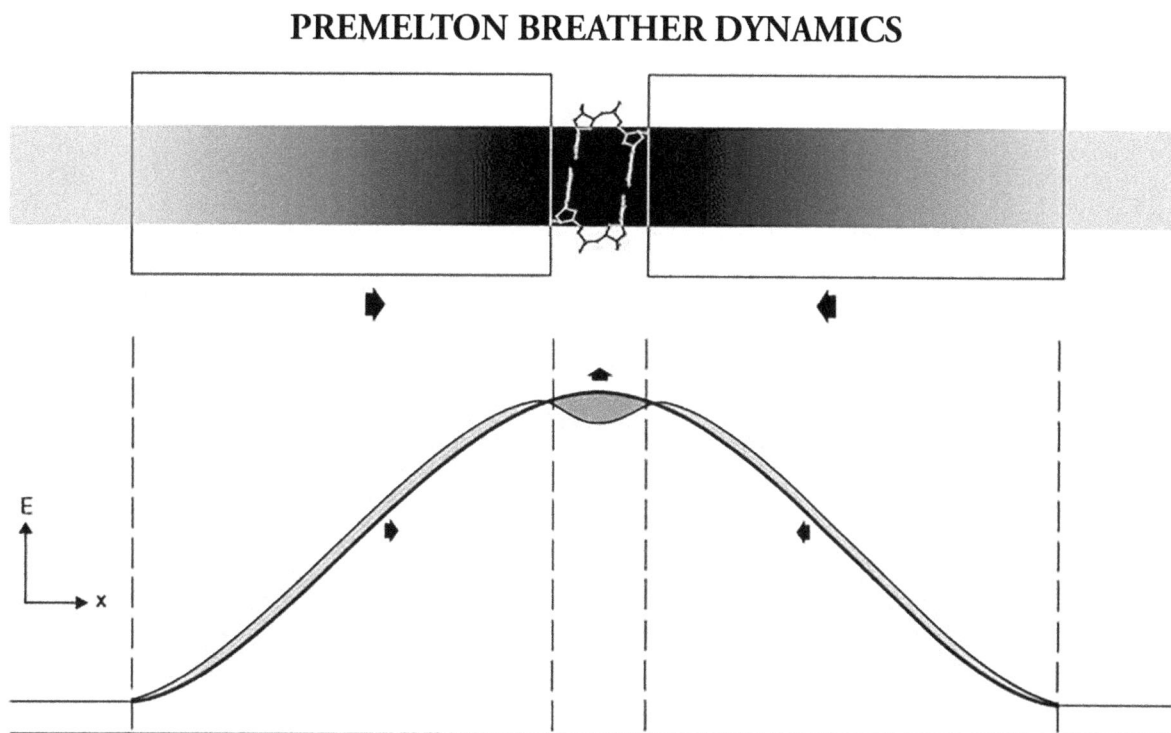

Figure 1: Lowest-amplitude breather-motion present within a premelton, showing the central beta-structural element alternating between its lowest- and highest-energy states (compare front- and back-panels). These hinge-like motions are coupled with the concerted movement of the kink and antikink boundaries (shown in the boxed-regions) on either side.

Isoenergetic breather-motions such as these demonstrate the collective-effect, an effect well known in many areas of physics. Small movements of atoms in sugar-residues within the kink and antikink boundaries combine together to give larger movements of atoms in base-pairs in the central beta-DNA core region (i.e., 0.025 Angstroms versus 2 to 3 Angstroms). This collective-effect explains how energy is transiently focused into the centers of premeltons to create an "open-state" into which drugs and dyes intercalate.

Larger-amplitude breather-motions (accompanied by the further stretching and unwinding of DNA) give rise to DNA-breathing distortions, which allow intercalators such as meso-tetra [4-N-methyl (pyridyl) porphine], echinomycin and nogalomycin to intercalate into DNA. They have not been shown in the figure above.

The molecular structures of premeltons in A- and B-DNA (i.e., B-B and A-A premeltons) -- as well as premeltons connecting A- with B-DNA (i.e., A-B and B-A premeltons) -- have been calculated by linked-atom least-squares methods. The results of these calculations are summarized in section 4, suggesting both classes of premeltons to span between 40 to 50 base- pairs.

As the kink and antikink move together for example, the energy-density in the central beta- DNA core region rises – this energy being used first to enhance (alternate) base-pair unstacking, and next to stretch and to eventually break hydrogen-bonds connecting base- pairs. As the kink and the antikink move apart, the reverse happens. Energy within this central core-region falls, allowing hydrogen-bonds to reform and base-pairs to partially restack.

PREMELTON BREATHER DYNAMICS

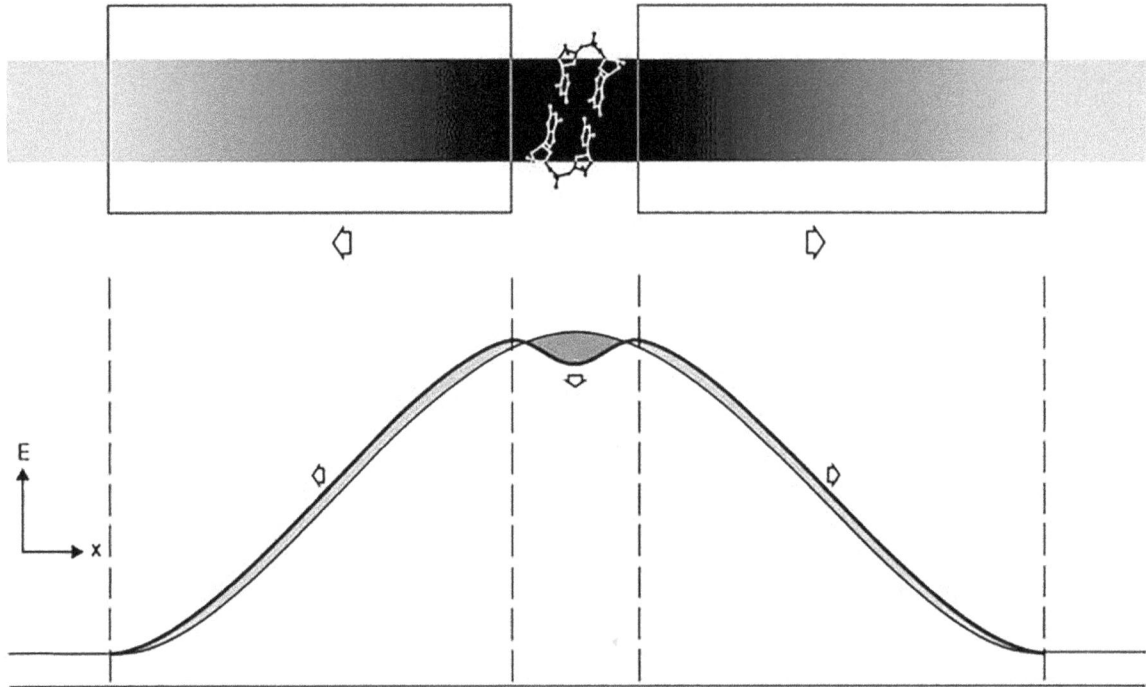

Figure 1: (continued).

[1] The terms "kink" and "antikink" have been used by both physicists and applied mathematicians to describe the solutions to a large-number of nonlinear partial differential- equations — they have precise meaning, being known as "topological-solitons". The "kink- antikink bound-state" on the other hand, represents a different class of solutions, these describing the emergence of coherent-structures that contain internal dynamical-motion (hence, the term, "breather-solitons", or, in lattice situations, "discrete-breathers"). Kink-antikink bound-states are encountered in a large number of diverse areas in nonlinear- science and are of particular interest to physicists and mathematicians working in these areas (readers unfamiliar with this area should consult references, 24-26).

[2] For the molecular biologist, the word "kink" has come to mean a sharp-bend in DNA due to a localized conformational-change in a sugar-residue and/or a phosphodiester- linkage. Although this terminology is somewhat restricted, it has proven useful in the DNA-area and poses no problem provided physicists and biologists agree on the meaning of the word "kink" in these two different contexts.

[3] Note in Figure 1, that movement in the kink- and the antikink-boundaries within premeltons is tightly coupled to the appearance of its lowest-and highest-energy states in the central beta- structural-element. The conformational extremes of these two-different energy-states, therefore, limit the excursions of the kink and antikink, causing them to remain together as a dynamical "kink-antikink bound-state". A qualitative discussion to understand how solitons originate in DNA to give rise to premeltons (and to meltons) with increasing-temperature can be found in the Addendum (section 10).

DNA BREATHING AND DRUG INTERCALATION ARE RELATED PHENOMENA

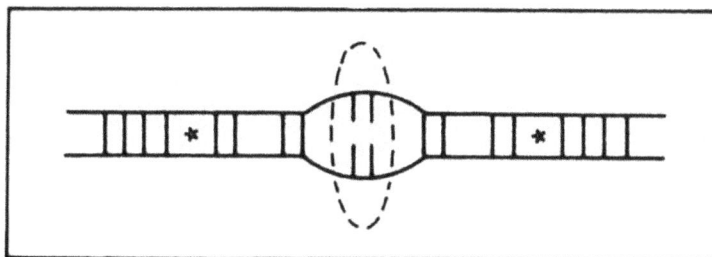

Figure 2 (Top): A simplified illustration showing DNA-breathing, a concerted dynamical process within premeltons that combines base-pair unstacking with the transient rupture of hydrogen-bonds connecting base-pairs. Premeltons arise spontaneously within these early- melting regions to nucleate DNA-melting, their central beta-structural core-regions providing a metastable DNA form which serves as an *activated-intermediate* in DNA-breathing, allowing planar drugs and dyes to intercalate into DNA. Base-pairs undergoing H-bond breakage are indicated in the dashed oval-area, while lower-energy beta-structural elements are indicated with asterisks on either side. Kink and antikink regions are not shown explicitly.

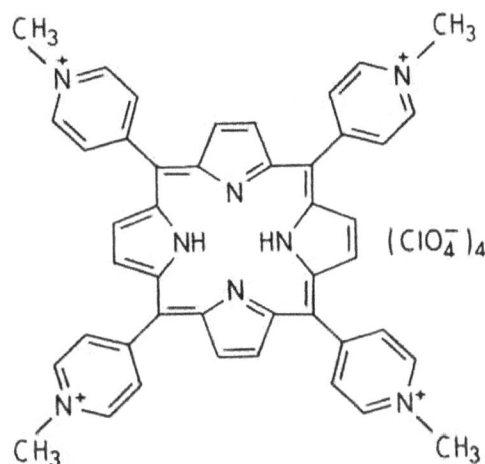

Meso-Tetra [4-N-methy (pyridyl) porphine]

Figure 2 (Middle): Intercalators that necessitate the transient rupture of hydrogen-bonds connecting base-pairs to gain entrance into (and exit out of) DNA (i.e., meso-tetra [4-N-methyl (pyridyl) porphine]) — constitute convincing evidence that DNA-breathing and drug- intercalation are related phenomena (35).

Isoenergetic breather motions such as these, facilitate the intercalation of drugs and dyes into DNA, and allow tritium exchange to occur at temperatures well below the melting temperature (33-34, 111-114). These motions demonstrate the collective effect, an effect well known in many areas of physics. Small movements of atoms in sugar- residues within kink and antikink boundaries combine together to give larger movements of atoms (within base-pairs) in the central beta- DNA core region (i.e., 0.025 Angstroms versus 2 to 3 Angstroms). This effect explains how energy is transiently focused into the centers of premeltons, to do the work required for DNA-breathing. See Figure 2.

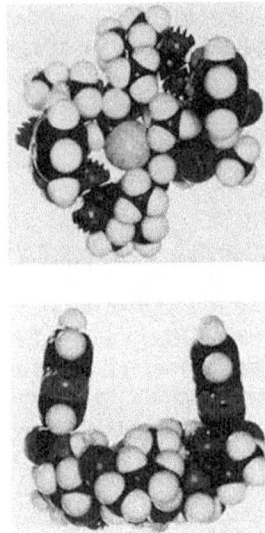

Figure 2 (Bottom): Echinomycin is an example of a bifunctional intercalator, having two quinoxaline rings separated by 10.2 Angstroms, connected through amide-linkages to a rigid octapeptide-chain. The stereochemistry of this naturally occurring DNA-binding antibiotic necessitates both quinoxaline ring-systems be able to intercalate simultaneously into neighboring high-energy beta-structural elements, and, for this reason, is a valuable probe to understand the detailed stereochemistry of DNA-breathing (36).

Premeltons nucleate DNA-melting and give rise to other types of DNA phase-transitions.To understand this, we begin by reviewing the structure and physical-properties of beta-DNA.

3. Beta-DNA, a Transition-State Intermediate in DNA Melting

Beta-DNA is a *distinctly different* structural form from either B- or A-DNA. Evidence for its existence comes from studies of intercalation by drugs and dyes, and by the binding of the *recA* protein to DNA (this information is summarized in the Appendix which follows). See Figure 3.

Although a duplex structure, beta-DNA is unique in being both *hyperflexible* and *metastable*. Its hyperflexibility suggests a semblance to a liquid-like phase, a phase in- between the more rigid B- and A- forms (i.e., both solid-like phases) and the melted single-stranded form (i.e., a gas-like phase). Both properties necessitate beta-DNA to be "pinned" by an intercalator or held by a protein, in order to be studied in detail.

The underlying source of nonlinearity separating beta-DNA from either A- or B- DNA, is the presence of alternating *C2' endo* and *C3' endo* sugar-pucker conformations along the sugar-phosphate backbone, combined with the partial-unstacking of alternate base-pairs[4]. The beta-structure is composed of repeating units, called beta-structural elements. These are a family of base-paired dinucleotide structures, each having the same mixed sugar-puckering pattern (i.e., C3' endo (3'-5') C2' endo) and having similar backbone conformational angles, but varying in their degree of base-unstacking. Lower-energy beta-structural elements contain base-pairs partially- unstacked, while higher-energy beta-structural elements contain base-pairs more completely-unstacked.

THE STRUCTURE AND PROPERTIES OF BETA-DNA

A-DNA
(C3′ *endo*)

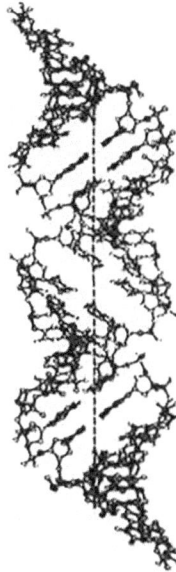

***β*-DNA
Alternating**
(C3′ *endo*-C2′ *endo*)

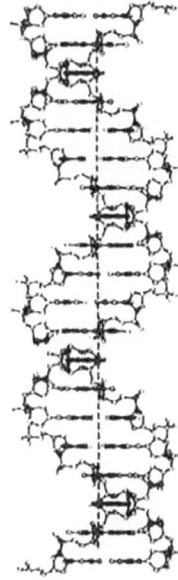

B-DNA
(C2′ *endo*)

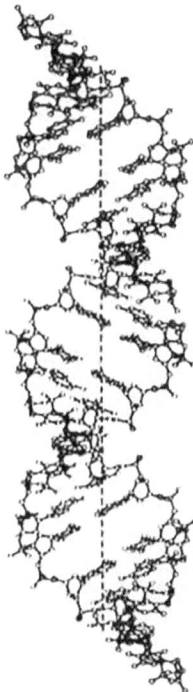

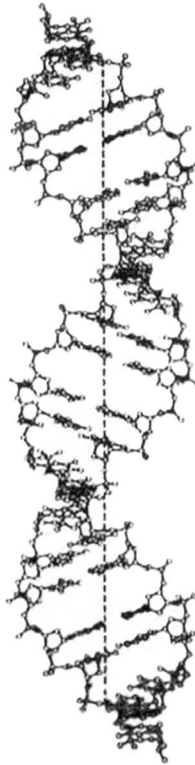

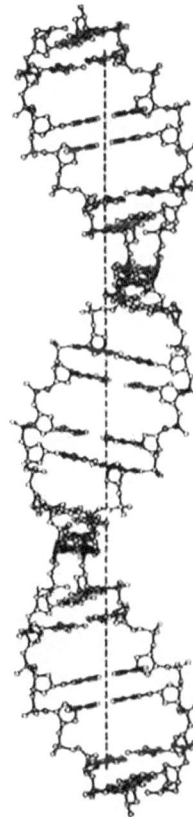

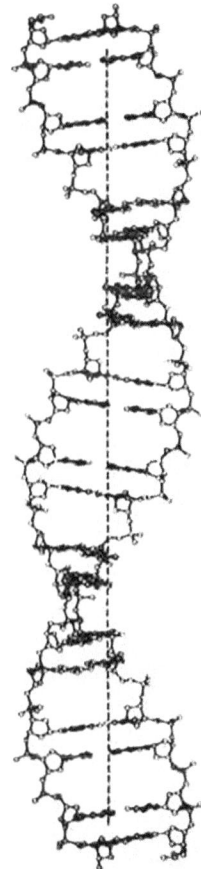

lowest-energy state   **highest-energy state**

Figure 3 (Top illustrations): A-DNA, beta-DNA and B-DNA, and their associated sugar-pucker conformations. The lowest-energy beta-DNA form is a metastable-structure, with helical-parameters intermediate between A- and B- forms. For comparative purposes, each figure contains 20 base-pairs.

Figure 3 (Bottom illustrations): Beta-DNA is also a hyperflexible-structure that exists in many different energy states. It is bounded on the left by its lowest-energy state and on the right by its highest-energy state. Steroidal-diamines such as irehdiamine A stabilize its lowest-energy state by partial-intercalation, while planar drugs and dyes, such as ethidium stabilize its highest- energy state by complete-intercalation. The lowest- energy state is also proposed to be a transition- state intermediate in the B- to A- transition, while the highest-energy state -- being a maximally extended and unwound DNA duplex-structure -- is proposed as being a transition- state intermediate in DNA-melting.

[4] **Deoxyribose sugar-residues, both as individual molecules or joined within the polymer structure, can assume either C2' and C3' endo pucker-conformations, both conformations having similar energies. Through the use of the pseudorotational-parameter (a mathematical parameter that defines the sugar-conformation), one can explore the energies of the complete range of conformational states. These calculations show energy minima at C2' and C3' endo regions, these being connected by a minimal-energy pathway having a barrier of about 1.5 Kcal/ mole. In B-DNA, sugar-residues have C2' endo puckers, whereas in A-DNA, they have C3' endo puckers. The transition-region separating these two sugar-pucker conformations is, therefore, a key source of nonlinearity that separates the A- and B- conformational states. Beta- DNA utilizes a similar source of nonlinearity (i.e., the beta-structural element contains both C3' endo and C2' endo sugar-puckers) to distinguish it from the A- and B- forms. Its metastability reflects the presence of additional energies in its structure that necessitate the partial-unstacking of alternate base-pairs in its lowest-energy form (31,32).**

Why is the (highest-energy) beta-DNA form a transition-state intermediate in DNA melting?

This is because beta-DNA lies along the *minimal-energy pathway* connecting B-DNA with single-stranded melted-DNA. Three distinctly different sources of nonlinearity appear as DNA-chains unwind, and these determine the sequence of conformational-changes that appear along this pathway. The first two sources of nonlinearity stem from changes in the sugar-pucker conformations and base-pair stacking. These require small- energies (i.e., ~kT), and appear as part of the initial structural-distortions accompanying DNA-unwinding. Starting with B-DNA, the effect of unwinding-DNA is to counter balance this with an equal, but opposite right-handed superhelical writhing to keep the linking invariant[5]. This is achieved through a modulated *beta-alternation* in sugar-puckering, accompanied by the gradual partial-unstacking of *alternate* base-pairs[6]. The lowest-energy beta-DNA form emerges as an end result, its metastability reflecting the presence of additional energies in its structure, giving rise to the partial-unstacking of alternate base-pairs.

[5] **Linking (L), Writhing (W) and Twisting (T) are related to each other by the relationship, L = W + T. Whereas L assumes only integral values, W and T can have either integral or fractional values (37-40).**

[6] **This dimerization phenomenon is known in solid-state physics as the Peierl's distortion and reflects the presence of spontaneous symmetry-breaking accompanying a nonlinear-exitation. See, for example (41,42).**

The third source of nonlinearity arises from the stretching and ultimate-rupture of hydrogen-bonds connecting base-pairs. At first, the beta-DNA structure is able to accommodate further unwinding through the gradual loss of superhelical writhing. This reflects the appearance of beta-structural elements having increasingly higher- energy, these having base-pairs more completely unstacked and unwound. Eventually however, a limit is reached and further unwinding begins to stretch and to break hydrogen-bonds connecting base-pairs. Duplex DNA therefore passes through a *transition state intermediate* (i.e. the highest-energy beta-DNA form), before attaining its single-stranded melted-DNA form[7].

[7] **It follows that single-stranded DNA (ssDNA) should possess structural features that resemble the beta-form. Evidence indicating this comes from electron-microscopic length measurements of both recA-dsDNA and recA-ssDNA filaments, which demonstrate DNA in both complexes to be stretched one and one-half times [43, 44]. Since synapsis between homologous DNA molecules begins with the invasion of a recA-ssDNA filamentous complex into an actively-breathing DNA-duplex region [45], DNA must be both complementary in sequence and structurally compatible for pairing to occur. The presence of a transient single-stranded beta-DNA like form within the premelton could, therefore, play a key role in initiating homologous-pairing with the recA-ssDNA complex in this early synaptic step required for synapsis in genetic recombination.**

Why is beta-DNA a metastable structural-phase?

As we have already said, this reflects the presence of additional-energies in its structure that necessitate beta-structural elements remain partially-unstacked. For example, partial base-pair unstacking may be necessary to relieve the strain-energy accumulating in the sugar-phosphate backbone (and/or elsewhere in the structure) as the lowest energy beta- form emerges (in the presence of heat and/or negative superhelicity) from either A- or B- DNA.

Why is beta-DNA a hyperflexible structural-phase?

This is because partial base-pair unstacking present within each beta-structural element weakens the van der Waals stacking-interactions that remain between base- pairs. Additionally, the ability of DNA to stretch and to unwind is enhanced, due to the presence of unhindered-rotation around glycosidic-bonds and other sugar- phosphate linkages. A combination of these two effects gives the beta-structural element hinge- like properties and for these reasons the beta-DNA form is expected to be a hyperflexible structural- phase.

We have proposed the highest energy beta-DNA form to be a transition-state intermediate in DNA-melting and in addition to act as a transition-state intermediate in the B- to A- structural phase-transition.

4. B- to A- Structural Phase Transition

Using the technique of linked-atom least-squares (46), it has been possible to compute the structural-intermediates that lie along a minimal-energy pathway connecting B - with A - DNA. This was accomplished by calculating a series of uniform transitions along the polymer, in which the sugar-puckering of alternate deoxyribose residues were altered incrementally and the structures then energy-minimized subject to a series of constraints and restraints[8].

MOLECULAR STRUCTURES OF THE B-A AND B-B PREMELTONS

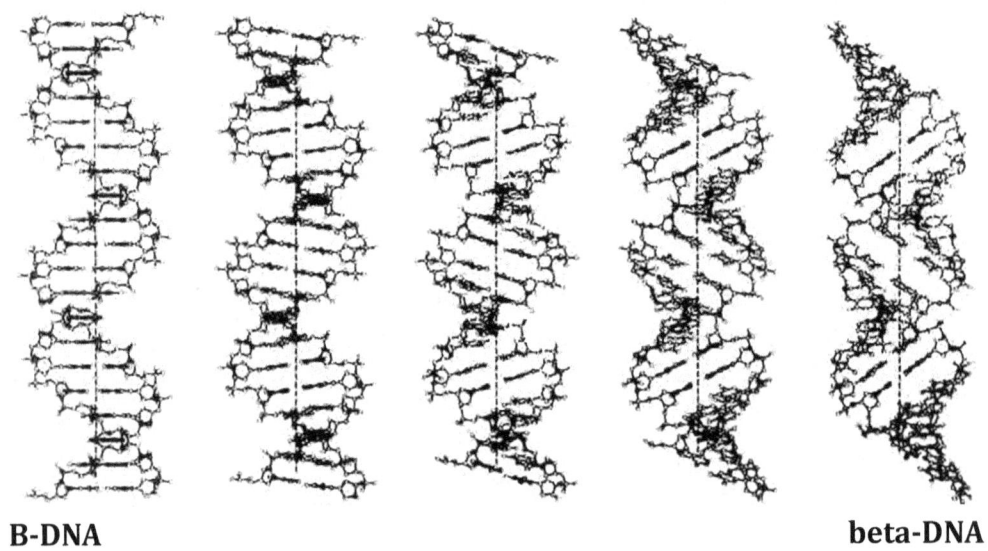

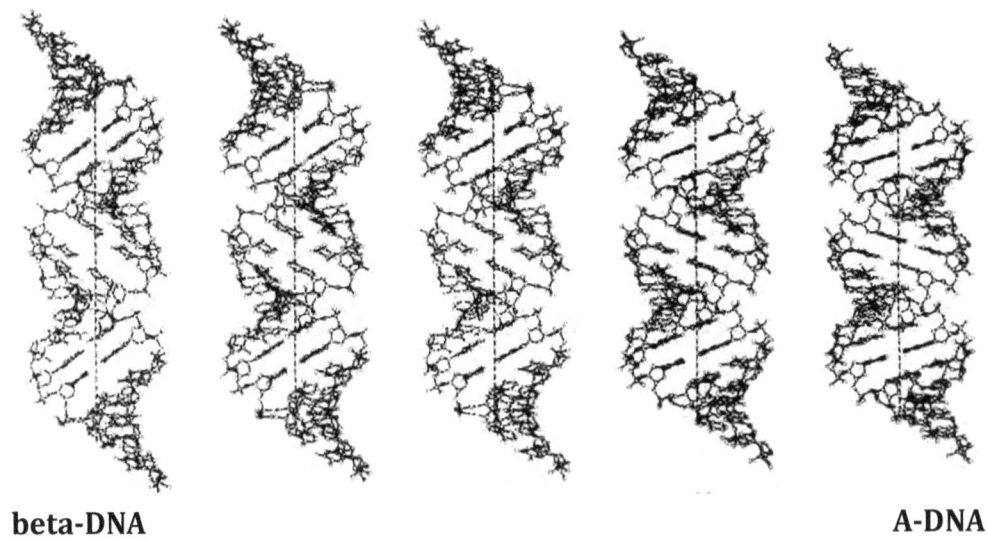

B-DNA   ←——————— B-A PREMELTON ———————→   A-DNA

B-DNA   ←——————— B-B PREMELTON ———————→   B-DNA

B-DNA   beta-DNA

beta-DNA   A-DNA

Figure 4 (Top illustrations): The molecular-structures of B-A and B-B premeltons. To simulate these structures, base-paired dinucleotide elements obtained from the modeling studies described below, were pieced together using a least-squares procedure. It is seen that, whereas B-A premeltons are (globally) *topological*, B-B premeltons are (globally) *nontopological*. The formation of these two different classes of premeltons demonstrates the concept of a bifurcation – this has led to a mechanism to understand how the B- to A- structural phase-transition takes place.

Figure 4 (Bottom illustrations): The molecular-structures connecting B- with A- DNA, computed as a uniform- transition along the polymer-chain by the method of linked-atom least- squares. Twenty-five structures have been calculated by this procedure although, for simplification, only nine have been shown here. Dinucleotide elements from each structure were then pieced together to form the premelton structures shown above.

[8] In these calculations, physicists will recognize sugar puckers to be the "masters", torsional angles defining the sugar-phosphate and sugar-base conformations, the"slaves".

In this way, we have discovered the existence of a minimal-energy pathway connecting B- with A-DNA which passes through its lowest-energy beta-DNA form. A detailed description of these calculations, together with the final-coordinates of all twenty-five structures, can be found in *"Kink-Antikink Bound States in DNA Structure" in Biological Macromolecules and Assemblies, Volume 2, Nucleic Acids and Inter-active Proteins, edited by F.A. Jurnack and A. McPherson, John Wiley & Sons, New York (1985)* (29, 31) [9].

To simulate the bound-states structures, base-paired dinucleotide elements from each structure were then pieced together using a least-squares procedure. The decision to compute twenty-five structural intermediates that connect B-DNA with A-DNA without having a detailed knowledge of the total-energies involved is somewhat arbitrary but does not alter the basic conclusions presented here.

[9] Detailed-calculations have shown that there is little or no base-pair unstacking in the first-half of the B-to beta-DNA (and the A- to beta-DNA) phase-boundaries. The combination of DNA-unwinding, counter- balanced by right-handed superhelical writhing, is achieved almost entirely, by "rolling" adjacent base- pairs (upon each- others van der Waals surfaces) towards the wide-groove direction, accompanied by a gradual- modification of (alternate) sugar-pucker geometries within the polymer- backbone. As one passes over the energy-barrier separating C2' endo from C3' endo sugar- conformations, there is a more abrupt-onset of partial base-pair unstacking to relieve the strain- energies in the sugar-phosphate chain that would otherwise develop. We have observed that it is necessary to relax the exact-requirement that only alternate-sugars are involved in the transition. To get over the energy-barriers arising in these intermediate-states, it is necessary to gently "rock" the other sugar-residue "backwards", towards the *C2' exo* conformation (in the B- to beta- pathway), or towards the *C3' exo* conformation (in the A- to beta- pathway) -- this readily allowing passage through these structural barriers.

It is seen that, whereas B-A premeltons are (globally) *topological*, B-B premeltons are (globally) *non-topological* – the formation of these two different classes of premeltons demonstrate the concept of a *bifurcation*, which leads to a mechanism to understand how the B- to A- structural phase-transition takes place [10].

[10] A bifurcation is defined as an event that takes place at a branch-point in a pathway to give rise to two-different outcomes. Although the source of the nonlinearity (i.e., that determines the pathway) remains the same, the decision as to which pathway to take at the branch-point is influenced by a bias. In the case of the B to A transition originating within the centers of premeltons, prevailing thermodynamic-conditions provide the bias.

Although B-A and A-B premeltons are moveable-boundaries, B-B and A-A premeltons are not, experiencing trapping- potentials that determine the likelihood they form within specific DNA-regions. It is evident that sequentially homogeneous DNA-polymers are not adequate models to understand the properties of naturally occurring DNA, because they lack sufficient information in their nucleotide base-sequence to give rise to this site- specificity. In the context of soliton-models, this poses the problem of describing soliton- behavior in the presence of locally-altered potentials. A general theory for this has been developed (47).

The theory shows that solitons either move nonuniformly or are trapped by locally favorable potentials. It remains to extend this theory to DNA structure to predict the formation of kink-antikink bound-states at specific nucleotide base-sequences. Because the kink-antikink bound-state is multiple base-pairs in extent (our calculations predict this number to be about fifty), it may be that the effective trapping-potential (this would depend on the energies necessary to partially unstack base-pairs in different DNA regions), involves the recognition of an extended sequence, rather than being determined by any single base-pair energetics, or its immediate neighbors.

The B to A structural phase-transition can now be understood in the following way. In the presence of suitable thermodynamic-conditions (i.e., those that create a bias that favors the formation of A-DNA), kink and antikink boundaries in B-B premeltons move- apart, forming longer and longer core-regions whose centers modulate into A-DNA. Such a mechanism is reversible and illustrates how a *bifurcation* within the central core- region of a premelton can lead to the B to A (or A to B) structural phase-transition. See Figure 4.

5. Higher Energy Kink-Antikink Bound States in DNA Structure

Bifurcations arising within premeltons having higher-energy (i.e., these containing larger beta DNA core-regions that undergo more vigorous breather-motions), can lead to the formation of two additional types of higher-energy kink-antikink bound-states in DNA. Called *branch-migratons* and *dislocaton-pairs*, each arises from different types of chain- slippage events, known as double- and single-stranded branch-migration (31).

See Figure 5.

Branch-migratons arise from transient breathing-distortions within premeltons, causing DNA chains to come apart and to then "snap-back", at nucleotide base-sequences having 2- fold symmetry (48). Weakly hydrogen-bonded hairpin-structures initially form, lengthened by a series of kinetic-steps in which hydrogen-bonds connecting base-pairs within dinucleotide-elements in vertical-stems break and simultaneously reform in horizontal-stems (or vice-versa), in a concerted 2-fold symmetric-process – this being called, "cruciform-extrusion" (49). A branch-migraton contains four-stems, each stem containing kink and antikink boundaries connecting beta-DNA core-regions with the surrounding B-DNA. In three-dimensions each stem is in all likelihood, pseudo-tetrahedrally coordinated (see Figures 6a and 6b)).

Dislocaton-pairs arise at repetitive base-sequences as, for example, in poly d (G-A): poly d (C-T) (50). Again, these structures form as the result of particularly energetic breathing events, causing DNA chains to come apart and to then "snap back" – forming small single- stranded "bubble-like" protrusions on opposite chains. These protrusions then move apart, leaving growing regions of beta-DNA in-between. The centers of these regions modulate into kink and antikink boundaries, and these in turn, continue to move apart to leave B- DNA. The net result is the appearance of dislocaton-pairs which move in opposite directions along DNA. Movement involves single-chain slippage and is remarkably similar to the mechanism underlying the movement of crystal-lattice dislocations – hence the term *dislocaton-pairs*. See Figure 7.

Breather-motions within branch-migratons facilitate double-stranded chain-slippage causing kink and antikink in horizontal stems to move in, while kink and antikink in vertical stems move out (and vice-versa) all in a concerted two-fold symmetric fashion. These motions require simultaneity of movement in both the horizontal- and vertical- stems, this following as a necessary consequence of the coherence present within this slippage-structure forming base-pairs in dinucleotides *within* beta-elements, rather than *between* beta-elements.

Low-amplitude breather-motions within a dislocaton-pair again require the dynamic- interchange between its lowest- and highest- energy beta-structural elements, coupled with the concerted movement of the kink and antikink boundaries on either side. Higher- amplitude breather-motions are expected to facilitate double-stranded chain-slippage events, these again requiring the breaking and rejoining of H-bonds connecting base-pairs in dinucleotides *within* beta-elements, rather than *between* beta-elements.

The formation of these two different kinds of higher-energy kink-antikink bound-state structures is another example of bifurcations emanating within the centers of premeltons. The underlying source of the nonlinearity that determines the path of the bifurcation, is the breaking and reforming (after chain-slippage) of hydrogen-bonds connecting dinucleotide base-pairs. The decision as to whether or not branch-migratons or dislocaton- pairs form is determined by the information coded in the nucleotide base- sequence – the sequence information constituting the bias.

The combined presence of torsional and writhing strain-energies found in negatively- superhelical circular-DNA increases the probability that branch-migratons and dislocaton- pairs arise at the appropriate sequences. These energies are first used to form premeltons. They are next used to form dislocaton-pairs or branch-migratons, and to propagate chain- slippage events. The energy in negatively-superhelical DNA can also be used to facilitate the B- to Z- structural phase-transition.

We will now discuss this.

6. B- to Z- Structural Phase-Transition

Although B- and A- DNA are right-handed double-helical structures, DNA molecules containing the alternating poly d (G-C): poly d (G-C) sequences under certain conditions, can assume a *left- handed double-helical conformation* (i.e., in the presence of high-salt and/or negative superhelicity) (51). This structure, called Z-DNA, contains the dinucleotide (G-C) as the asymmetric-unit held together by Watson-Crick base-pairs. Being a left-handed double-helical structure, Z-DNA contains sugar-phosphate backbone conformations radically-different from either B- or A-DNA [11].

FORMATION OF THE BRANCH-MIGRATON AND DISLOCATON-PAIRS

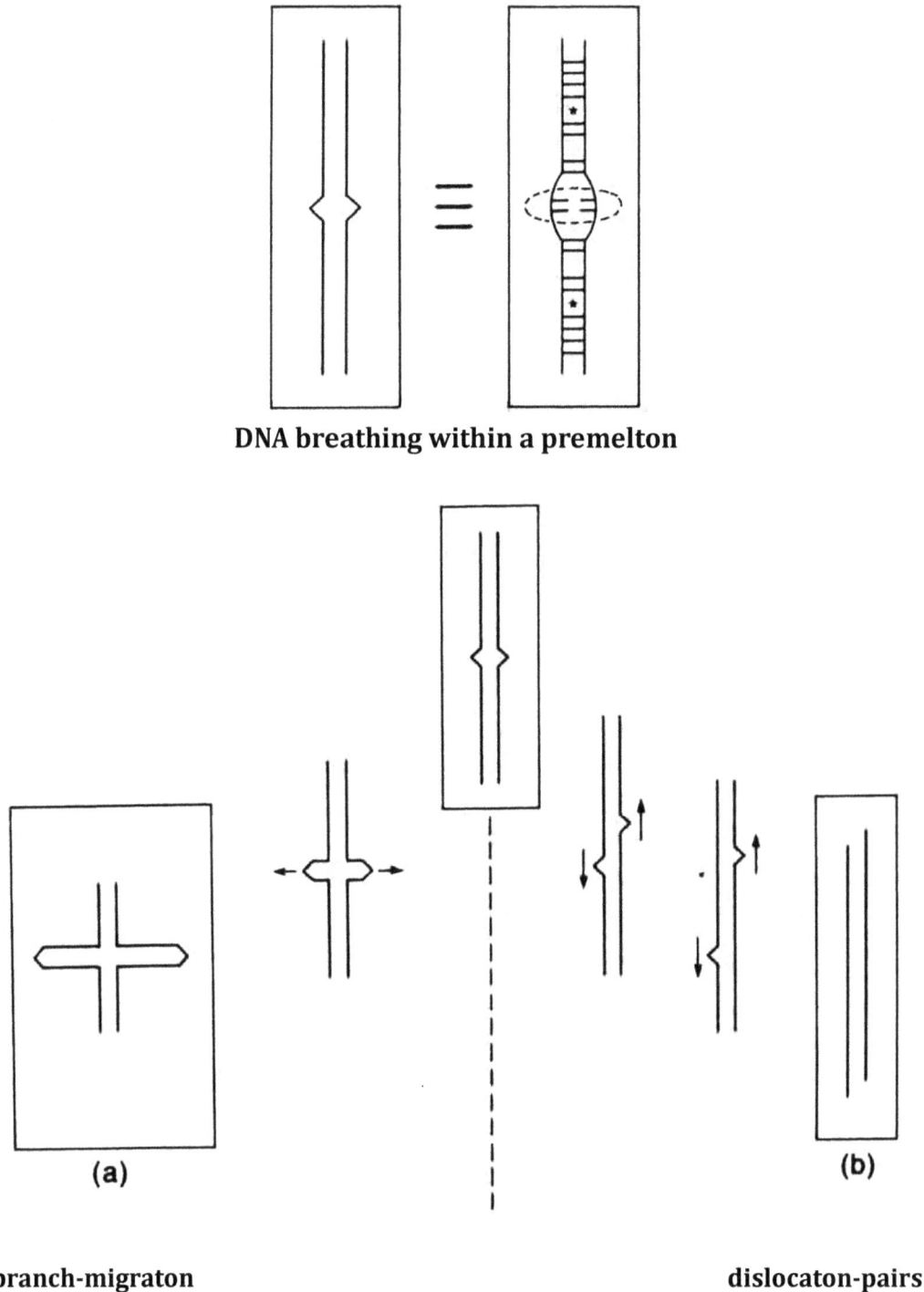

DNA breathing within a premelton

(a)   **(b)**

branch-migraton   **dislocaton-pairs**

Figure 5: A possible mechanism interrelating the origin of different types of chain-slippage events that occur in DNA. Both types of chain-slippage events are facilitated by the presence of negative-superhelicity in DNA.

BRANCH MIGRATON BREATHER DYNAMICS

Figure 6 (a): Low-amplitude breather-motions in the branch-migraton, showing the dynamic- interchange between the lowest- and highest-energy beta-structural elements coupled with the concerted movement of the kink and antikink boundaries on either side. Higher- amplitude breather-motions are expected to facilitate double-stranded chain-slippage events, these requiring the breaking and rejoining of H-bonds connecting base-pairs -- they have not been shown in this figure.

BRANCH MIGRATON BREATHER DYNAMICS

Figure 6 (b). See text for discussion.

[11] Major differences include deoxyguanosine in the *syn* conformation with sugars *C3' endo*, and deoxycytidines in the anti-conformation with sugars C2' endo (i.e., syn and anti refer to the two different orientations of the purine or pyrimidine bases relative to the sugars and are determined by the torsional angles around the glycosidic-bond, chi). The twist angle relating adjacent base-pairs in CpG steps is -8 degrees, while for GpC steps it is -52 degrees - this being associated with large-changes in the sugar- phosphate conformation. Finally, dyad-axes fall between base-pairs, but not at the level of each base-pair, since the asymmetric-unit in Z-DNA is a base-paired dinucleotide.

DISLOCATON BREATHER DYNAMICS

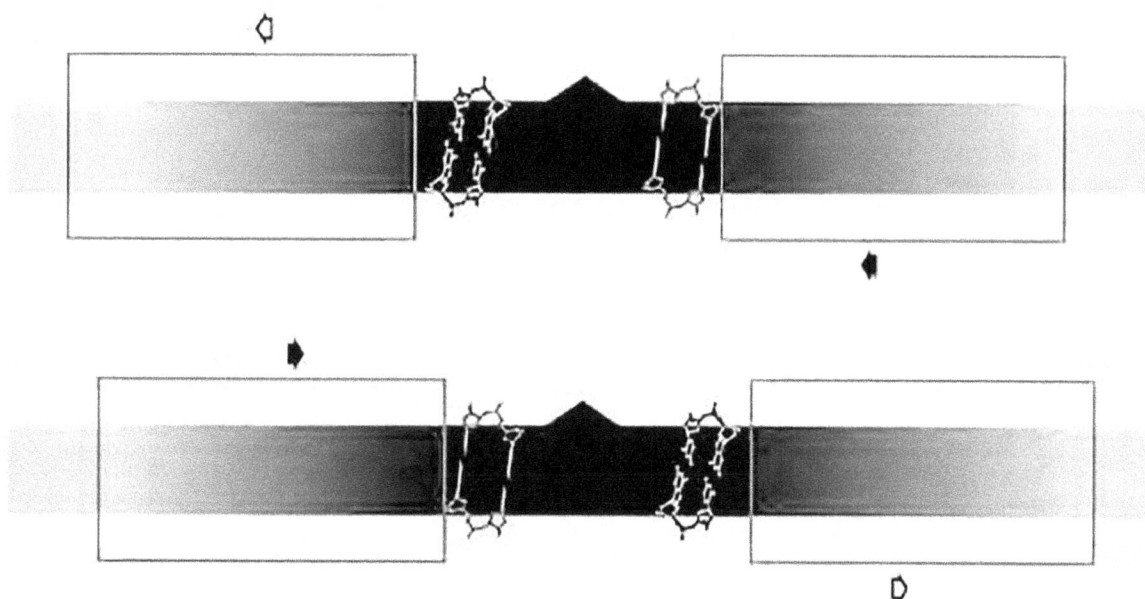

Figure 7: Low-amplitude breather-motions within a dislocation-pair showing the dynamic-interchange between its lowest- and highest- energy beta-structural elements coupled with the concerted movement of the kink and antikink boundaries on either side. Higher-amplitude breather-motions are expected to facilitate double-stranded chain-slippage events, this requiring the breaking and rejoining of H-bonds connecting base-pairs in the highest- energy beta- structural element – this has not been shown in this figure.

The B- to Z- transition begins as the result of a bifurcation that arises of during the formation of dislocaton-pairs. Initially, actively breathing premeltons form within B- DNA, which in the presence of high-salt and/or negatively superhelicity, transform spontaneously into dislocaton-pairs. The molecular boundaries that allow the helix to "swing left" capitalize on the additional-flexibility and length of single-stranded DNA regions on opposite chains. Both B-Z boundaries form in a concerted 2-fold symmetric fashion, this being a direct consequence of the nonlinearity that ties the process together. However, due to its extreme complexity, it is not possible to say anything more about the stereochemistry of the B-Z dislocaton-pairs.

A fundamental prediction of the model however, is the appearance of single-stranded DNA-regions juxtaposed to beta-DNA regions, which, in turn are juxtaposed to B-DNA through kink and antikink boundaries. The model predicts S1-nuclease sensitive-sites to lie within these junctions and predicts high-cooperativity to accompany ethidium- binding (leading to increasing lengths of beta-DNA "pinned" by ethidium at the expense of Z-DNA), both predictions being supported by existing experimental evidence (52-54)[12]. See Figures 8, 9, and 10.

[12] **Related observations have been reported by Jaroslav Kypr and Micheala Vorlickova for the steroidal-diamine, dipyrandium, a molecule similar to irehdiamine A, also known to partially intercalate into DNA (55-61). In the presence of increasing concentrations of dipyrandium, poly d(G-C) has been shown to switch from its left-handed Z form into its right-handed neighbor- exclusion form (i.e., the lowest-energy beta-DNA form), as followed by circular-dichroism studies in the ultraviolet spectral-region (62).**

A MECHANISM TO UNDERSTAND THE B- TO Z- TRANSITION

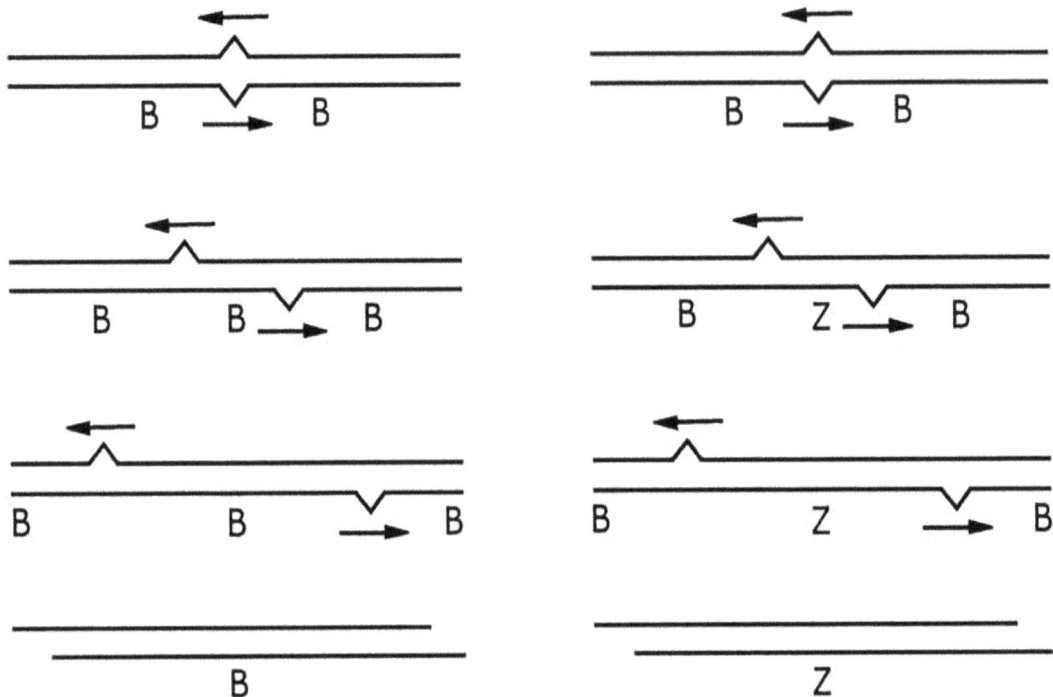

Figure 8: (Left illustrations) B-B dislocaton-pairs form from actively-breathing premeltons within B- DNA regions, which then move apart leaving growing B-DNA regions behind. The model predicts single chain-slippage events to involve an even-number of nucleotide-bases. This sequence of events can also occur with A-DNA or A-RNA containing the appropriate (even-numbered) repetitive base-sequences.

Figure 8: (Right illustrations) B-Z dislocaton-pairs form from actively-breathing premeltons within B-DNA regions, which then move apart leaving growing Z-DNA regions behind. Again, the model predicts single chain-slippage events to involve an even-number of nucleotide- bases. This sequence of events can also occur in A-DNA or A-RNA, containing the appropriate (even- numbered) repetitive base-sequences.

ETHIDIUM BROMIDE AS A COOPERATIVE EFFECTOR OF A DNA STRUCTURE

Pohl FM, TM Jovin, W Baehr, JJ Holbrook. *Proc Nat Acad Sci* 1972; 69; 3805-3809

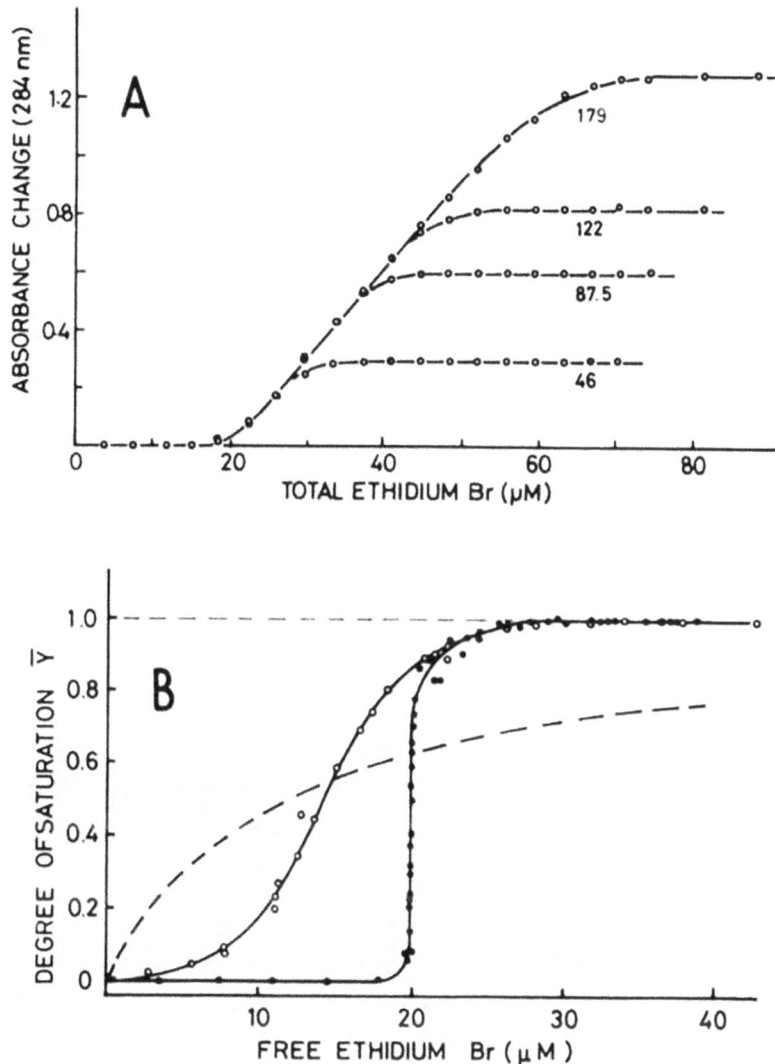

Figure 9: Binding of ethidium to the low- and high- salt forms of poly (dG-dC). (A) Absorbance change (284 nm) versus total ethidium bromide added to different concentrations of DNA solutions (B) degree of saturation versus free ethidium in solution for the low-salt form (dashed lines) of poly (dG-dC) 100-200, high-salt form (open circles) of poly (dG-dC) 10, and high-salt form (closed circles) of poly (dG-dC) 100-200 (reproduced, with permission, from Pohl, F. et al, 1972 (53)).

A MECHANISM TO UNDERSTAND THE COOPERATIVE BINDING BY ETHIDIUM TO THE ALTERNATING d (G-C) POLYMER UNDER HIGH SALT CONDITIONS

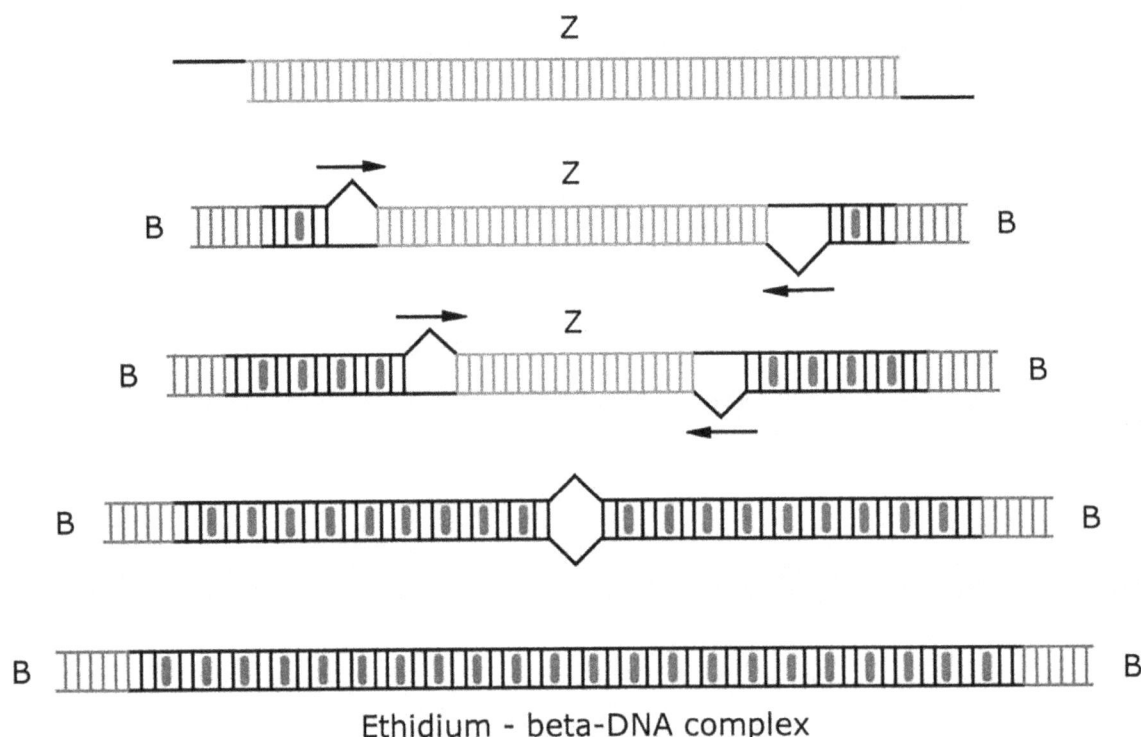

Ethidium - beta-DNA complex

Figure 10: The alternating poly d (G-C) synthetic polymer undergoes the B- to Z- transition under high-salt conditions. Under these conditions, ethidium demonstrates high- cooperativity when binding to the polymer (53, 54). What is the mechanism underlying this cooperativity?

The data indicate the emergence of a new structural-phase, this new phase being the ethidium: (highest- energy) beta-DNA complex. The cooperative nature of the binding-reaction is a direct consequence of the presence of the beta-DNA phase-boundaries connecting Z-DNA with B-DNA, and their ease of movement. The phase-transition begins with ethidium-binding to small beta- DNA regions located within dislocaton-pairs surrounding Z-DNA ("nucleation", see top illustrations). This is followed by the sequential-binding of ethidium into expanding- domains of beta-DNA, at the expense of Z-DNA ("propagation", see middle illustrations). Finally, extensive- regions of the neighbor-exclusion ethidium: *highest-energy* beta-DNA complex form that replaces Z-DNA (see bottom illustrations). This mechanism is most easily understood as being a structural phase-transition, in which the ethidium: highest-energy beta-DNA complex emerges as the dominant-phase. See Figure 9, which shows the ethidium- poly d (G-C) cooperative- binding isotherm observed under high-salt conditions.

An analogous model can be used to understand the A- to Z- RNA transition, which makes similar predictions.

7. Implications in Molecular Biology

A. *Premeltons in DNA*

Soliton-concepts provide a strong rationale for expecting coherent nonlinear-excitations to extend over multiple base-pairs, to provide either transient or permanent conformational- changes in specific DNA-regions. These will always be present to a certain extent at normal thermal-energies – however, their concentrations will depend on temperature, pH, ionic- strength, hydration, extent of negative-superhelicity and other thermodynamic factors.

The stability of a premelton is further expected to reflect the collective properties of nucleotide base-sequences in extended DNA-regions. Since the ease with which beta- structural elements form is correlated with the magnitude of localized base-stacking energies, base-sequences with minimal base-overlap (i.e., as occur, for example, in alternating pyrimidine-purine sequences) are favored, along with sequences that contain high A-T/G-C base-ratios. The energetics in the kink and antikink regions are another important factor. The tendency for a premelton to localize within a given DNA region would depend on the depth of the energy-minimum in the central core-region, coupled with the height and separation of the energy domain-walls (i.e., that are associated with the kink and the antikink structures) on either side.

B. *Mechanism of DNA Melting*

Accordingly, premeltons tend to localize at early melting-regions in DNA, and at elevated temperatures, serve to nucleate the melting-process. At lower temperatures, kink and antikink pairs (herringbone-regions) surround small beta-DNA core-regions (see cross- hatched regions). As the temperature rises, these bounding kink- antikink pairs move- apart, leaving growing beta-DNA cores, whose inner-regions begin to experience the nonlinear- stretching of hydrogen-bonds connecting base-pairs (stippled-regions). Finally, at still higher temperatures, single-stranded denaturation-bubbles appear, separated from regions of B- DNA (or A-DNA) by the complex phase boundaries already described. Such composite structures correspond to higher-energy structural solitons, and have been called, *meltons* (30). See Figure 11.

C. *Nuclease-hypersensitive sites in DNA*

The presence of premeltons undergoing low-amplitude breather-motions within promoter regions can serve the important function of providing nucleation-centers for site-specific DNA-melting by the RNA-polymerase enzyme, and perhaps, by other proteins necessary for gene-activation.

Key evidence for the existence of premeltons in eukaryotic promoter-regions has been provided by the discovery that the chemical-nuclease, 1, 10- phenanthroline-copper (I), an intercalator (63), cleaves DNA at hypersensitive-sites in naked relaxed circular DNA- molecules which contain the histone gene-cluster of *D. melanogaster*, many of these being located at the 5'- ends of genes. Remarkably, this small molecule has also been shown to mimic the larger micrococcal-nuclease in recognizing these sites, cleaving them with equal- frequencies. Pancreatic-DNase (I) is also known to cleave these same sites (with similar frequencies), as well as other unrelated-sites (64).

A MECHANISM TO UNDERSTAND DNA-MELTING

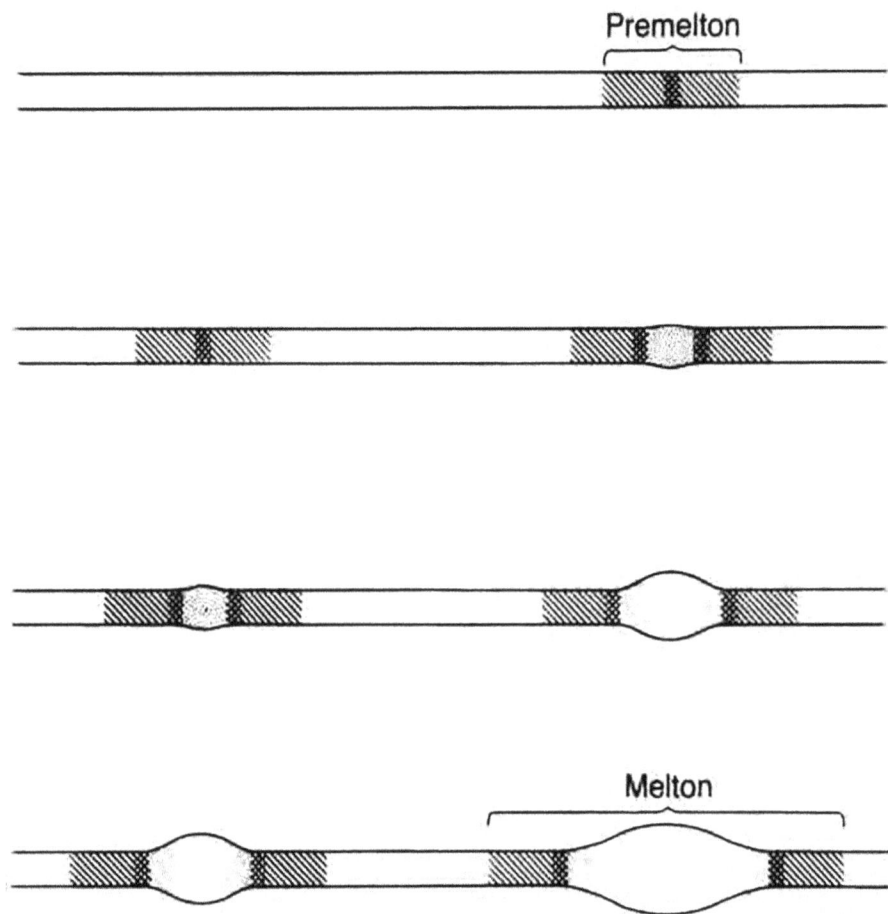

Figure 11: A schematic illustration of DNA-melting showing how premeltons become meltons with increasing temperature — these being examples of structural-solitons in DNA.

The addition of the single-strand specific DNA-binding protein to these same DNA- molecules made superhelical selectively melts DNA at many of these sites, as revealed by S1-nuclease studies in combination with electron-microscopy (65). Important additional information has been provided by studies of this same gene cluster in active- chromatin, where the micrococcal- nuclease cleaves hypersensitive-sites at the 5'-ends of genes, while the pancreatic-DNase (I) cleaves hypersensitive-sites at both the 5'- and 3'-ends of genes (66). Related observations have been made in the heat-shock locus 67B1 of D. melanogaster, where micrococcal-nuclease and 1, 10- phenanthroline-copper(I) have been found to cleave DNA at the 5'-ends of genes in both naked DNA and DNA in active-chromatin (67, 68).

What determines the degree to which a given DNA site is hypersensitive to nuclease- action?

The central beta-structural element within premeltons is proposed to be the substrate recognized and cleaved by these agents. As already discussed, premeltons arising in specific regions with higher-probability and existing for longer-lifetimes will be cleaved more frequently than those arising elsewhere.

Low-amplitude breather-motions within the premelton cause the central beta- structural element to alternate between its two-different energy-states. The highest- energy state (i.e., the highest-energy beta-structural element) is proposed to be the substrate cleaved by the micrococcal-nuclease and 1, 10- phenanthroline-copper (I) and the lowest-energy state (i.e., the lowest-energy beta-structural element), the substrate cleaved by the pancreatic-DNase (I).

Premeltons containing lower-energies are expected to undergo smaller breather- motions. These premeltons may not have enough energy to open the central beta- structural element wide- enough to act as a substrate for 1, 10- phenanthroline-copper(I) or the micrococcal- nuclease. Premeltons such as these, however, can remain as substrates for the pancreatic- DNase (I). This would explain the wider range of hypersensitive-sites recognized and cleaved by this enzyme, when compared with 1, 10- phenanthroline-copper (I) and the micrococcal- nuclease. See Figure 12.

D. *RNA Polymerase Binding to the Promoter*

The concept that premeltons, existing at the 5'- ends of genes, serve to initiate site-specific DNA melting by the RNA polymerase to begin transcription – while premeltons reforming at the 3'- ends of genes, facilitate the detachment of the RNA polymerase from DNA to terminate transcription allows one to understand additional key aspects of DNA transcription. See Figure 13.

E. *How Actinomycin Binds to DNA and Exerts its Mechanism of Action*

Actinomycin has been proposed to intercalate into beta-DNA, found within the boundaries connecting double-stranded B-DNA with single-stranded DNA in the transcription-complex (30). This immobilizes (i.e., "pins") the complex, interfering with the elongation of growing RNA-chains. In extremely active-genes, RNA polymerases lie in a close-packed arrangement along DNA. Interference with the movement of one polymerase by actinomycin is expected to affect the movement of other polymerases. This can explain why nucleolar RNA synthesis is so sensitive to the presence of actinomycin D (69-71). See Figure 14.

It should be noted that the mechanism above *necessarily predicts* premeltons to be at the 5'- ends of genes, facilitating the attachment of the RNA polymerase to its promoter, leading to site- specific melting. It also predicts premeltons to reform at the 3'- ends of genes, promoting the detachment of the RNA polymerase from its terminator, ending transcription. These predictions follow from earlier discussions of DNA-melting and its implications in understanding DNA transcription (refer to Figures 11-14).

F. *How Drugs and Dyes Intercalate into DNA*

Intercalators can be divided into two general classes – simple and complex.

1. *Simple Intercalators*

Ethidium is an example of a simple intercalator, binding to the beta-structural element in both DNA and RNA model dinucleotide systems. Its interactions are "simple" in the sense that it binds to the beta-structural element – utilizing stacking interactions with the base-pairs and electrostatic interactions with the sugar-phosphate chains.

Mechanistically — how does ethidium intercalate into DNA?

We have proposed the following mechanism:

1) Simple intercalators (i.e., ethidium) begin by "pinning" premeltons that arise spontaneously in DNA (i.e., at the micrococcal-nuclease hypersensitive- sites).

2) As the drug/DNA binding-ratio increases, ethidium continues to bind to DNA sequentially forming expanding domains of neighbor-exclusion structure, surrounded by moving kink- antikink phase-boundaries on either side ("propagation").

3) At saturation, practically all the DNA is complexed with ethidium, DNA being held in an elongated and unwound (highest-energy) beta-DNA form (see Figure 15).

4) The above process is most easily understood as being a *structural phase-transition,* in which the ethidium: beta-DNA complex emerges as the dominant phase.

A similar mechanism can be used to understand how simple intercalators bind to double- stranded RNA.

An important prediction of this model is that extended microcrystalline domains of neighbor-exclusion binding exists at high drug/DNA binding ratios. This prediction has been confirmed by fiber-diffraction studies, which indicate the platinum organometallo-intercalator, 2-hydroxy-ethane thiolate (2,2',2'' terpyridine) platinum (II) to form extended microcrystalline domains when it complexes with calf-thymus DNA (72) (see Figure16).

2. *Complex Intercalators*

Complex intercalators (for example, actinomycin, echinomycin and daunomycin) utilize additional "complex" types of interactions when binding to DNA. As a result of these interactions there may or may not be alterations in the sugar-phosphate geometries (with the concomitant variations in the magnitude of angular unwinding), sequence binding affinities and neighbor- exclusion volumes when binding to DMA. Their detailed binding mechanisms will differ depending on on the nature of the interactions utilized by each complex intercalator; nevertheless, the models below predict special features that each intercalator demonstrates when intercalating into premeltons arising in DNA.

Actinomycin, for example, has two cyclic pentapeptide chains related by 2-fold symmetry connected together by a phenoxazone-ring system. The phenoxazone ring intercalates into DNA, while the two pentapeptide-chains lie in the narrow-groove, forming specific hydrogen-bonds between the carbonyl-oxygens on both threonine residues and the 2-amino-groups on both guanine-residues (when binding to d-pGpC sequences) (73-77). More generally, actinomycin demonstrates a sequence-binding preference, binding to sequences of the type: d-pGpX, where X = C, T, G, or A. The steric bulk of the pentapeptide-chains increase its neighbor-exclusion volume to four nucleotide base-pairs, a value observed when actinomycin binds to poly d-(G-C) at high drug/DNA binding ratios. This means that, if actinomycin binds to the beta-DNA form, only every *other* beta-structural element is available for intercalation by the phenoxazone ring-system. Finally, X-ray crystallographic studies of actinomycin complexed to a number of different self-complementary oligonucleotides have confirmed the overall features of the original DNA-binding model (78-81). Echinomycin, on the other hand, is a bifunctional-intercalator, having two quinoxaline ring-systems separated by 10.2 Angstroms, connected through amide-linkages to a rigid octapeptide chain. The stereochemistry of this naturally occurring DNA- binding antibiotic necessitates that both quinoxaline ring-systems be able to intercalate simultaneously into neighboring high-energy beta-structural elements within the premelton. X-ray crystallographic studies of echinomycin complexed to d-GCGTACGC have indicated the existence of Hoogsteen base-pairing in the center of an unwound and elongated (highest- energy) beta-DNA like form, suggesting DNA to first come apart allowing the central TA region to isomerize before echinomycin stabilizes this structure. For this reason, echinomycin could be a valuable probe to understand the nature of structural rearrangements arising from DNA- breathing events (82-84). Echinomycin is a potent inhibitor of DNA transcription, its mechanism of action being similar to that of actinomycin D.

Daunomycin, as a final example, demonstrates intercalation by the aglycone anthra- cycline chromophore between terminal d-CpG base-paired dinucleotides (in the d-CpGpTpApCpG complex), the anthracycline chromophore being oriented at right-angles to the long-dimension of the base-pairs. The cyclohexene-ring rests in the minor-groove of the hexanucleotide duplex, its substituents hydrogen-bonding to base-pairs above and below each intercalation- site and interacting with phosphate-groups on opposite chains. All deoxyribose sugar-residues are in their *C2' endo* conformation – interestingly, however, no unwinding exists at the immediate intercalation-site. In this respect, the intercalation stereochemistry of daunomycin is entirely different from other complex- intercalators that have been studied (85-87).

So - how do such complex intercalators bind to DNA?

1) Actinomycin and echinomycin begin by intercalating into premeltons that arise spontaneously in DNA (i.e., the micrococcal-nuclease hypersensitive sites) – echinomycin requiring at least two beta-structural elements within an actively- breathing premelton for bifunctional-intercalation to come about.

2) Being complex intercalators, actinomycin and echinomycin possess in addition, nucleotide base-sequence binding preferences which, after being intercalated – influence their binding affinities to any particular DNA-site. This affects the final stability each complex has when binding to any given micrococcal-nuclease hypersensitive binding-site in DNA.

3) The presence of both pentapeptide-chains on actinomycin and the octapeptide-chain on echinomycin, may play a key additional-role in determining the affinity each molecule has to bind neighboring-sites. The presence of significant peptide-peptide interactions between neighboring-molecules could create cooperatively in the binding reaction, this amplifying or overriding the role sequence binding-preferences play in determining their interactions within any given binding-site in DNA.

4) As described previously, daunomycin intercalates anomalously into DNA. All deoxyribose sugar-residues remain in their C2' *endo* conformation and it is for this reason that there is stretching, but no unwinding at the immediate intercalation-site. This anomaly can easily be explained in the following way. Daunomycin begins by intercalating into a premelton in DNA – however (having its anthracycline-ring oriented at right-angles to the long dimension of the base-pairs) – lacks adequate stacking- energies to "pin" this premelton. The premelton spontaneously disappears, leaving daunomycin attached to a (distorted) B-type binding-site,

ANALOGOUS CLEAVAGE OF DNA BY MICROCOCCAL-NUCLEASE AND A 1, 10- PHENANTHROLINE-CUPROUS COMPLEX

Jessee B, G Gargiulo, F Razvi, A Worcel, *Nucleic Acids Research* 1982;
10 Number 19: 5823-5834

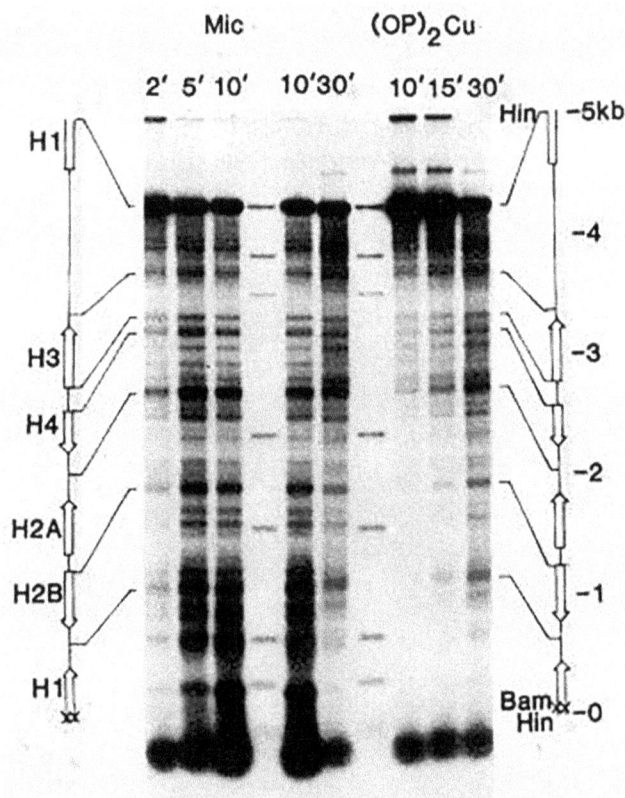

Figure 12: This study demonstrates the remarkable similarities between the chemical- nuclease, 1, 10- phenanthroline-copper (I), an intercalator, and the micrococcal-nuclease -- in their ability to recognize and to cleave hypersensitive-sites in a 5,000 base-pair circular-DNA fragment containing the histone gene-cluster from *D. melanogaster* (64).

The figure above shows a comparison between the micrococcal-nuclease and 1, 10- phenanthroline-copper (I) cleavage-patterns using agarose-gel electrophoresis followed by autoradiography. Circularized naked DNA- molecules -- previously labeled with radioactive phosphorous at a single Bam H1 site – were incubated with either the micrococcal-nuclease or 1, 10- phenanthroline-copper (I) and the reaction followed as a function of time. The resulting fragments were then cleaved with Hind III to give fragments having a common Hind III end, this being 68 base-pairs downstream from the labeled Bam site. Slab gel electrophoresis in 1% agarose followed by autoradiography, was then used to visualize radioactively-labeled fragments containing different DNA chain-lengths.

Cleavage patterns exhibited by both agents are amazingly similar, most hypersensitive-sites being found at the 5'-ends of genes or lying between adjacent genes. What is even more remarkable is the observation in subsequent experiments that many of these same sites nucleate melting when the single-strand specific DNA-binding protein of *E. coli* is added to these same circular DNA-molecules, made negatively-superhelical. The location of these small melted- DNA regions has been established using the S1-nuclease, in combination with electron- microscopy (65, 66).

A MECHANISM TO UNDERSTAND THE FORMATION OF THE RNA POLYMERASE-PROMOTER TIGHT BINDING COMPLEX

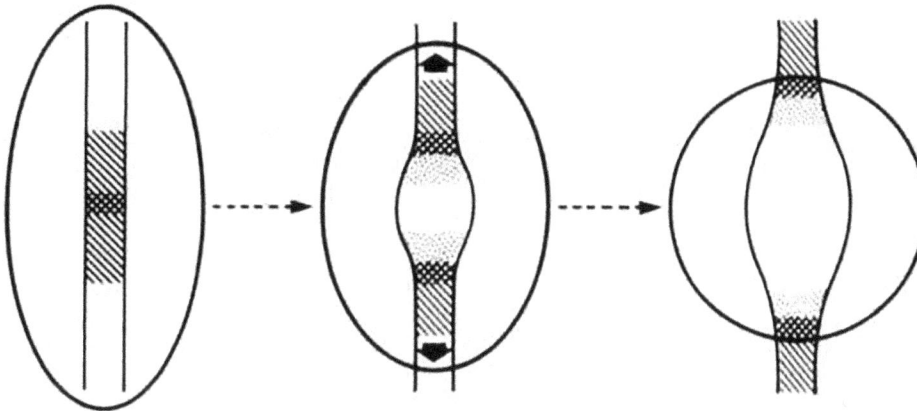

Figure 13: One can envision the formation of the transcriptionally-competent tight-binding complex to involve the initial-attachment of the polymerase to a premelton located at or near the start of transcription (shown on the left), triggering a cascade of conformational-changes in both the polymerase and the DNA (shown in the middle), that lead to the formation of the tight- binding complex (shown on the right). The process described above can be considered to be a series of concerted allosteric-transitions leading to the progressive-union of two molecular- species. How might this occur, and what is its underlying energetics?

This is best understood as being a protein-DNA structural phase-transition, the emergent- phase being the RNA- polymerase: promoter tight-binding complex. Complex-formation entails a series of stepwise conformational- transitions, in which energy is transferred from the polymerase to the DNA in the form of small packets (i.e., "kinks"). This is possible, providing the protein begins by being in a high-energy metastable-state. It can then spontaneously fall into lower-lying metastable-states, as DNA-melting and tight complex- formation ensue. Such an adiabatic process is expected to have little (or no) net-change in free- energy.

The mechanism is reversible. One can imagine the transcriptionally-competent tight-binding complex (shown on the right) to undergo a series of concerted allosteric-transitions (shown in the middle) that lead to the final detachment of the polymerase from the premelton (shown on the left). Such a mechanism necessarily accompanies the termination of transcription at the 3' ends of genes. The level of negatively-superhelical strain-energy in DNA provides the bias that determines the direction of this protein-DNA phase-transition. Other more active- processes can be involved as well.

It is well known that the transcriptionally-competent RNA-polymerase: DNA-complex is associated with an extremely large (apparent) binding-constant. Classical thermodynamics would predict a large net negative free-energy change to accompany the binding-reaction. If this were true, how then is it possible for the RNA- polymerase to move along DNA during the process of DNA-transcription?

This is understood in the following way.

The binding by the RNA-polymerase to the promoter is an adiabatic-process, energy being transferred from the protein to the DNA in a series of stepwise allosteric-transitions that lead to the formation of the transcriptionally-competent tight-binding complex (see above). Although there is little or no net free-energy change expected for such a process (this being an example of a protein-DNA structural phase-transition), the final structure contains both molecular species

ACTINOMYCIN AND DNA TRANSCRIPTION

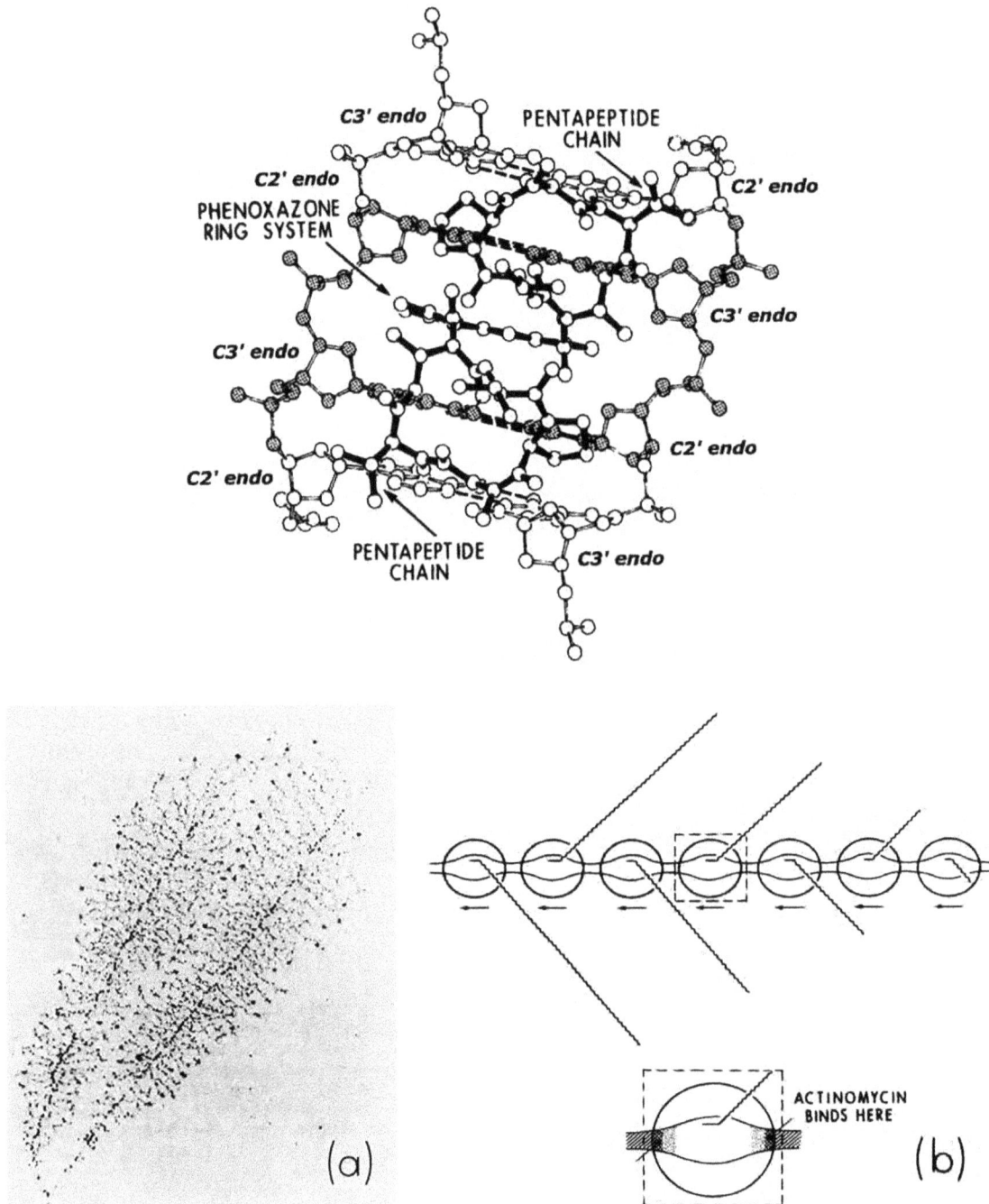

Figure 14: (Top illustration): Stereochemistry of actinomycin-DNA binding.

Figure 14: (Bottom illustrations): (a) Electron-micrograph of nucleolar genes undergoing transcription (69). (b) Interpretation of the micrograph in (a), showing the mechanism of action of actinomycin (30). Actinomycin binds to beta-DNA, a conformational-intermediate that exists within the boundaries connecting double-stranded B-DNA (or A-DNA) with single- stranded DNA in the transcription-complex. This immobilizes the complex, interfering with the elongation of growing RNA-chains [13, 14].

topologically linked together (i.e., in a way analogous to how two oppositely oriented "easy-zippers" are connected together, when becoming attached to the tracks on a "Ziploc" plastic bag). Such a model predicts the transcription-complex to be able to "slide" with minimal friction along DNA during transcription, in spite of the large apparent binding-constant holding these molecular-species together. This model accounts for the processivity observed in RNA synthesis as well.

The tight-binding transcriptionally competent complex arises as the result of topological linking (i.e.,"intertwining"), *not* from the presence of a large negative free-energy change accompanying complex formation. The extremely large (apparent) binding-constant for this complex, therefore, should not be confused with a true equilibrium binding constant, as described by classical thermodynamics.

[13] Leroy Liu and James Wang have provided a key insight into the nature of DNA supercoiling accompanying DNA transcription that has shed-light on this question (70). They have theorized that in the presence of significant resistance to the rotational motion of the RNA polymerase and its nascent RNA-chain around DNA during transcription, the advancing polymerase generates positive superhelicity in the DNA template ahead of it, and negative superhelicity behind it. In nucleolar genes, where it is estimated that there may be as many as 200 RNA polymerases moving-down the DNA template while synthesizing growing ribosomal RNA chains, positive and negative superhelical DNA- regions between transcription complexes annihilate one another causing adjacent transcription complexes to bond-together to form "trains" of transcription complexes, these now moving synchronously along DNA. If this were the case, then the binding by one actinomycin molecule is sufficient to stop the entire "transcription-train" from moving along DNA.

[14] Earlier, Harold Weintraub and Mark Groudine discovered that active-genes were particularly sensitive to pancreatic DNase (I), the degree of sensitivity being correlated with the level of gene-activity (71). Their findings are compatible with active genes being in the beta-DNA form, not only within each transcription complex, but also between adjacent transcription- complexes. This could explain why active genes are extremely sensitive, both to the presence of actinomycin D and to the action of pancreatic DNase(I).

which has a stereochemistry allowing substituents on the cyclohexene ring to more completely interact with adjacent base-pairs and phosphate-groups on opposite chains. Of course – to maintain microscopic reversibility – it is necessary for daunomycin to await the formation of another premelton at this site, at which time it can readily dissociate from DNA.

8. Experimental Predictions.

What definitive experiments can be done to test the model?

The model suggests the following experiments.

1) A key prediction of the model is that premeltons delineate the 5'- and 3'- ends of genes from the noncoding DNA regions surrounding them. A pilot project with naked T7 DNA (for example) can be carried out to determine whether this is true or not.

This can be done in two separate ways.

STRUCTURE OF THE ETHIDIUM-DNA NEIGHBOR
EXCLUSION BINDING MODEL

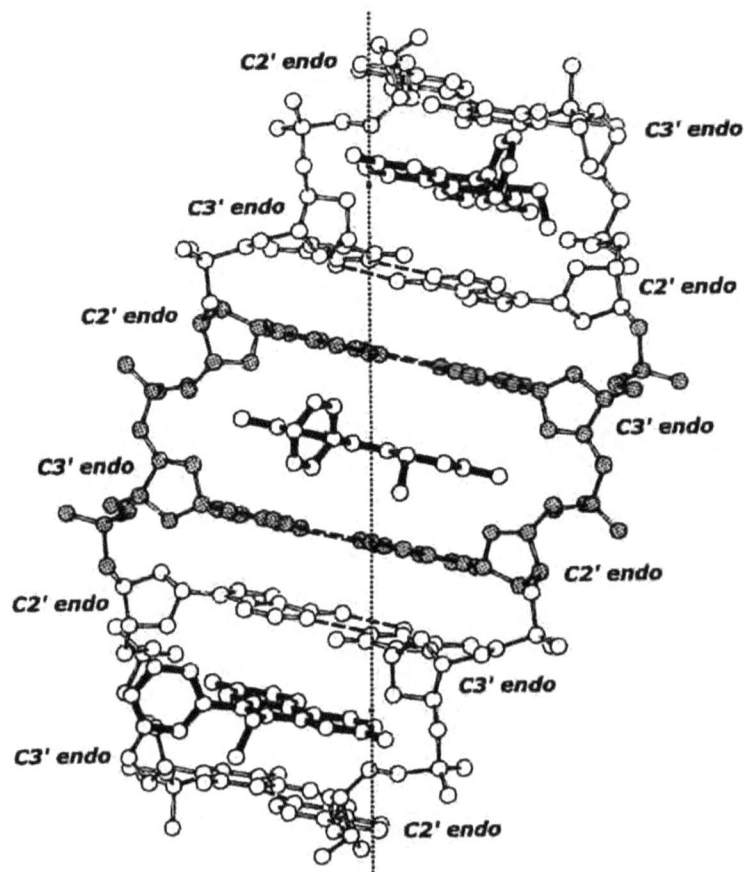

Figure 15: Structure of the ethidium: DNA neighbor-exclusion binding model. See text for discussion.

[15] It has been known for many years that the upper-binding limit for intercalators into naturally DNA (and RNA) is one intercalator bound every 2.2 to 2.3 base-pairs. This can be explained by intercalation beginning at the centers of premeltons (i.e., at the micrococcal- nuclease hypersensitive-sites), these being separated by an odd or even number of base-pairs with equal-probability. "Dead regions" are expected to appear (i.e. regions where there is no intercalation), due to an interference effect between approaching kink and antikink regions that are out of phase with one-another. A similar situation can also exist in the B- to A- phase transition, where small regions of B-DNA are expected to remain even in the presence of thermodynamic conditions favoring the A-form (see earlier discussion describing this).

[16] Concerning the lack of cooperativity observed when simple-intercalators bind to DNA -- this probably reflects the large number of nucleation-events occurring at the micrococcal- nuclease hypersensitive-sites. This would tend to hide the true sigmoidal- nature of the binding- isotherm, and to give one appearing to be purely hyperbolic. Further experiments should be carried out to examine this question.

X-RAY FIBER DIFFRACTION EVIDENCE FOR NEIGHBOR-EXCLUSION BINDING OF A PLATINUM METALLO INTERCALATION REAGENT TO DNA

Bond PJ, R Langridge, KW Jennette, SJ Lippard, Proc. Natl. Acad. Sci. USA 1975; 72: 4825-4829

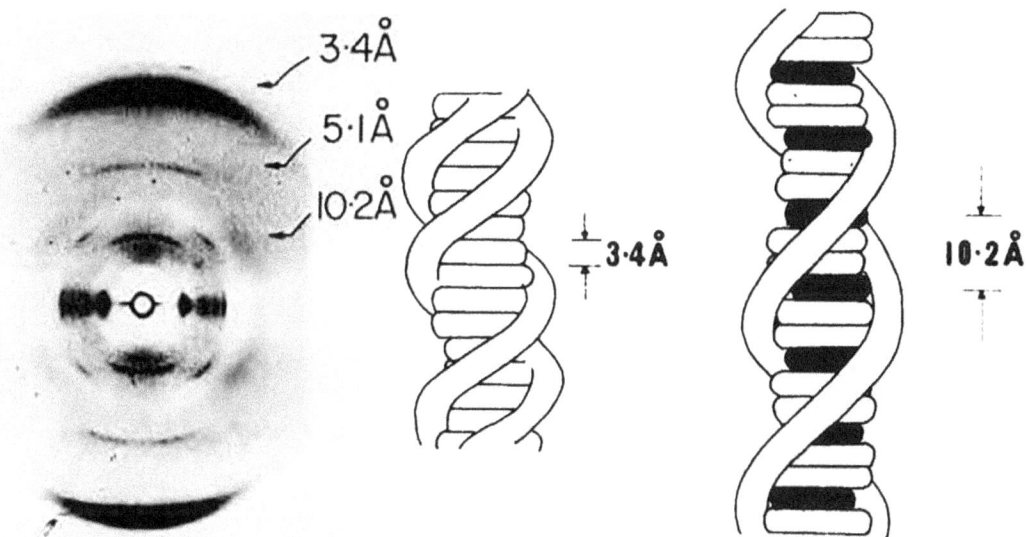

Figure 16: An X-ray diffraction pattern obtained from polycrystalline fibers, containing 2-hydroxyethane- thiolato (2, 2', 2''-terpyridine) platinum (II) bound to calf- thymus DNA by intercalation. This fiber-diffraction study confirms the presence of neighbor-exclusion binding by this platinum organometallo- intercalator, when it binds to DNA at high drug/DNA binding-ratios. From the sharpness of the 10.2 and 5.1- Angstrom reflections (these primarily reflecting the platinum-platinum scattering- vectors), it is possible to estimate the crystalline domain-size to be in the order of several-hundred (intercalated) base-pairs. These data are most easily understood as indicating the presence of a structural phase-transition, in which the platinum organometallo-intercalator: (highest-energy) beta- DNA complex emerges as the dominant- phase.

First -- my model predicts that the 5'- ends of genes are hypersensitive to *both* the micrococcal-nuclease and the pancreatic-DNase (I), whereas the 3'- ends of genes remain hypersensitive to only the pancreatic-DNase (I).

This is because premeltons at the 5'- ends of genes undergo 'breather motions', which cause their central-regions to alternate between their *highest- energy* beta- structural element (i.e., the substrate for the micrococcal-nuclease), and their *lowest-energy* beta-structural element (i.e., the substrate for the pancreatic- DNase (I)). Premeltons at the 3'- ends, on the other hand, lack adequate-energies to undergo significant breather-motions, and for this reason, are only sensitive to the action of the pancreatic-DNase enzyme, which recognizes and cleaves the lowest- energy beta-structural element in their centers.

Second -- my model predicts that, at the 5'- ends of genes, *both* irehdiamine and ethidium competitively inhibit the micrococcal-nuclease and pancreatic-DNase (I), whereas, at the 3'-ends, only irehdiamine (but *not* ethidium) competitively inhibits the pancreatic-DNase (I).

2) The model further predicts ethidium (but *not* irehdiamine) to competitively inhibit the micrococcal-nuclease from cleaving hypersensitive-sites that exist *between* nucleosomes in whole chromatin. It also further predicts irehdiamine (but *not* ethidium) to competitively inhibit the pancreatic-DNase (I) from cleaving hypersensitive-sites *within* nucleosomes.

3) The model predicts micrococcal nuclease and pancreatic-DNase (I)) hyper-sensitive-sites be *present* in long-chain DNA-molecules, but *absent* in shorter- chain oligonucleotide duplexes.

The histone gene cluster in *D. melanogaster* could be an ideal model-system to study this. The general plan would be to synthesize a series of DNA-molecules containing the micrococcal-nuclease hypersensitive-site (located at, say, the 5'- end of the H4 gene), surrounded by increasing-lengths of the histone gene-cluster sequence on either side. Should a gradual enhancement in sensitivity to cleavage by the micrococcal-nuclease take place, this can be detected by agarose-gel electrophoresis. Ideally, two new fragments will appear – these can be purified, and, if needed, their nucleotide-sequence determined.

Such an experiment would allow one to determine whether (or not) a nuclease hypersensitive-site requires the presence of a more extended-structure to be on either side. A premelton is expected to span a minimum of about fifty base-pairs from end to end. To observe a micrococcal-nuclease hypersensitive-site, therefore, it may be necessary to create a DNA-fragment one-hundred or more base-pairs long.

4) The existence of the high-energy beta-DNA (or -RNA) form "pinned" by ethidium (or propidium) can be established by determining the structure of a suitable complex by X-ray crystallography – for example, propidium complexed to d-TACGTA, and to r- UACGUA. The structure determination of the lowest-energy beta-DNA (or - RNA) form "pinned" by irehdiamine can also be determined – for example, irehdiamine complexed to d-TACGTA, and to r-UACGUA. Finally, higher- resolution data of the actinomycin-oligonucleotide complexes can be recollected, and the structures solved by Takusagawa and his colleagues reexamined, in order to establish the detailed DNA-conformation present in these complexes.

9. Appendix: Further evidence indicating the existence of beta-DNA

A. *The Discovery of the Beta-Structural Element: Simple versus Complex Intercalators*

There is currently a wealth of crystallographic information from model drug-nucleic acid crystal-line-complexes that has been contributed by our laboratory, along with other laboratories around the world. These studies have indicated that intercalators fall into two separate classes – simple and complex. Here, we focus on the discovery that simple intercalators exclusively bind to the beta-structural element, when complexed with a series of self-complementary dinucleoside-mo-nophosphates. *Simple intercalators (shown in Figure 17 and in Table 1) bind to the beta-structural element in these DNA- and RNA- model dinucleotide systems.* Their interactions are "simple" in that they exclusively utilize stacking-interactions with the base-pairs and electrostatic interactions with the sugar- phosphate chains (88-101).

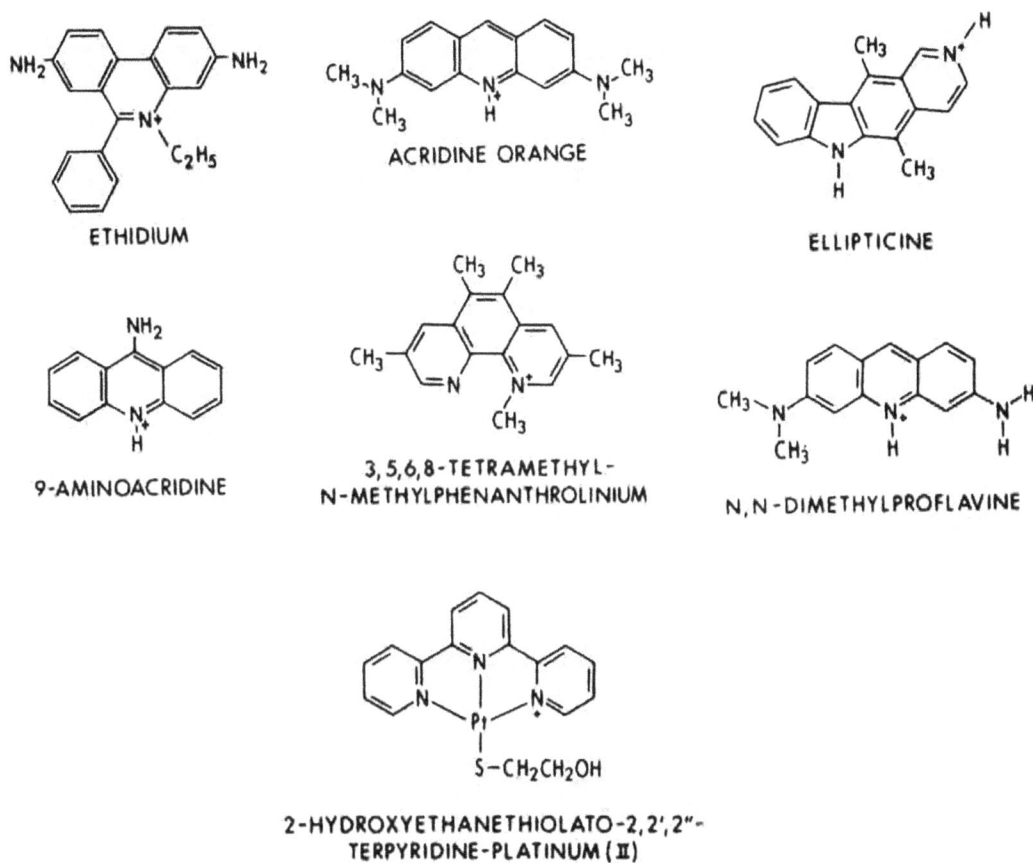

Figure 17: Chemical structures of simple intercalators.

Simple intercalators demonstrate a sequence binding preference CpG and TpA in these model solid-state studies and in related solution studies (102, 103). This reflects the ease of unstacking pyrimidine-purine sequences versus purine-pyrimidine sequences, an important consideration in understanding how simple-intercalators bind to synthetic DNA- (and RNA-) like polymers such as poly d (A-T) and poly d (G-C). Other effects predominate however, when simple intercalators bind to long chain naturally occurring DNA molecules (see later discussions).

A typical study that demonstrates their binding to the beta-structural element is ethidium: cytidylyl (3'- 5') guanosine, shown in Figure 18 (97). This complex consists of an intercalated ethidium-molecule (shown with dark covalent bonds) and a stacked ethidium-molecule (shown with light covalent bonds) above and below the intercalated complex. Sugar-phosphate conformations contain the mixed sugar-puckering pattern: cytidylyl C3' *endo* (3'-5') C2' *endo* guanosine.

The beta-structural element has been observed in 15 separate crystallographic determinations. These involve seven different intercalators complexed to a variety of DNA-like and RNA-like self-complementary dinucleoside monophosphates (see Table1). *Four* structures (5-iodocytidylyl (3'-5') guanosine complexed to ellipticine, acridine- orange, tetramethyl-N- methylphenanthrolinium, and N, N-dimethylproflavine) are *isomorphous*, and therefore demonstrate a *host-guest relationship*. The remaining 11 structures crystallize in different lattice environments that contain variable-numbers of water-molecules.

Table 1. Unit-cell Constants, Space Groups, and Structural Data for Simple Intercalators Complexed to DNA- and RNA-like, Self-complementary Dinucleoside Monophosphates

Complex	St[a]	Cell constants	Space group	Sugar puckering	Twist angle	Reference
Ethidium: 5-iodouridylyl(3'-5') adenosine	2:2	$a = 28.45$ Å $b = 13.54$ Å $c = 34.13$ Å $\beta = 98.6°$	C2	C3'-endo (3'-5') C2'-endo	$8 \pm 1°$	Tsai et al. (1977)
Ethidium: 5-iodocytidylyl(3'-5') guanosine	2:2	$a = 14.06$ Å $b = 32.34$ Å $c = 16.53$ Å $\beta = 117.8°$	P2₁	C3'-endo (3'-5') C2'-endo	$8 \pm 1°$	Jain et al. (1977)
Ethidium: uridylyl(3'-5') adenosine	2:2	$a = 13.70$ Å $b = 31.67$ Å $c = 15.13$ Å $\beta = 113.9°$	P2₁	C3'-endo (3'-5') C2'-endo	$9 \pm 1°$	Jain and Sobell (1982a)
Ethidium: cytidylyl(3'-5') guanosine	2:2	$a = 13.79$ Å $b = 31.94$ Å $c = 15.66$ Å $\beta = 117.5°$	P2₁	C3'-endo (3'-5') C2'-endo	$9 \pm 1°$	Jain and Sobell (1982b)
Ethidium: cytidylyl(3'-5') guanosine	2:2	$a = 13.64$ Å $b = 32.16$ Å $c = 14.93$ Å $\beta = 114.8°$	P2₁	C3'-endo (3'-5') C2'-endo	$8 \pm 1°$	Jain and Sobell (1982b)
Acridine orange: 5-iodocytidylyl(3'-5') guanosine	2:2	$a = 14.33$ Å $b = 19.68$ Å $c = 20.67$ Å $\beta = 102.1°$	P2₁	C3'-endo (3'-5') C2'-endo	$10 \pm 1°$	Reddy et al. (1979)
Acridine orange: cytidylyl(3'-5') guanosine	1:2	$a = 12.65$ Å $b = 11.79$ Å $c = 15.74$ Å $\alpha = 92.7°$ $\beta = 107.0°$ $\gamma = 90.0°$	P1	C3'-endo (3'-5') C2'-endo	$9 \pm 1°$	Wang et al. (1979)
Ellipticine: 5-iodocytidylyl(3'-5') guanosine	2:2	$a = 13.88$ Å $b = 19.11$ Å $c = 21.42$ Å $\beta = 105.4°$	P2₁	C3'-endo (3'-5') C2'-endo	$11 \pm 1°$	Jain et al. (1979)
3,5,6,8,-Tetramethyl-N-methyl phenanthro-linium: 5-iodocytidyl (3'-5')guanosine	2:2	$a = 13.99$ Å $b = 19.11$ Å $c = 21.31$ Å $\beta = 104.9°$	P2₁	C3'-endo (3'-5') C2'-endo	$11 \pm 1°$	Jain et al. (1979)
9-Aminoacridine: 5-iodocytidylyl(3'-5') guanosine	4:4	$a = 13.98$ Å $b = 30.58$ Å $c = 22.47$ Å $\beta = 113.9°$	P2₁	C3'-endo (3'-5') C2'-endo C3'-endo (3'-5') C2'-endo	$8 \pm 2°$ $10 \pm 2°$	Sakore et al. (1979)
N,N-Dimethylproflavine: 5-iodocytidylyl(3'-5') guanosine	2:2	$a = 11.78$ Å $b = 14.55$ Å $c = 15.50$ Å $\alpha = 89.2°$ $\beta = 86.2°$ $\gamma = 96.4°$	P1	C3'-endo (3'-5') C2'-endo	$9 \pm 1°$	Bhandary et al. (1982)
N,N-Dimethylproflavine: 5-iodocytidylyl(3'-5') guanosine	2:2	$a = 14.20$ Å $b = 18.99$ Å $c = 20.73$ Å $\beta = 103.6°$	P2₁	C3'-endo (3'-5') C2'-endo	$9 \pm 1°$	Bhandary et al. (1982)
N,N-Dimethylproflavine: cytidylyl(3'-5') guanosine	2:2	$a = 14.44$ Å $b = 19.62$ Å $c = 18.22$ Å $\beta = 95.0°$	P2₁	C3'-endo (3'-5') C2'-endo	$9 \pm 1°$	Sakore et al. (1982)
N,N-Dimethylproflavine: deoxycytidylyl(3'-5') deoxyguanosine	2:2	$a = 20.79$ Å $b = 33.82$ Å $c = 13.40$ Å	P2₁2₁2	C3'-endo (3'-5') C2'-endo	$9 \pm 1°$	Sakore et al. (1982)
Terpyridine platinum: deoxycytidylyl(3'-5') deoxyguanosine	2:2	$a = 22.25$ Å $b = 13.57$ Å $c = 33.12$ Å	P2₁2₁2₁	C3'-endo (3'-5') C2'-endo	$12 \pm 1°$	Wang et al. (1978)

[a] Stoichiometry.

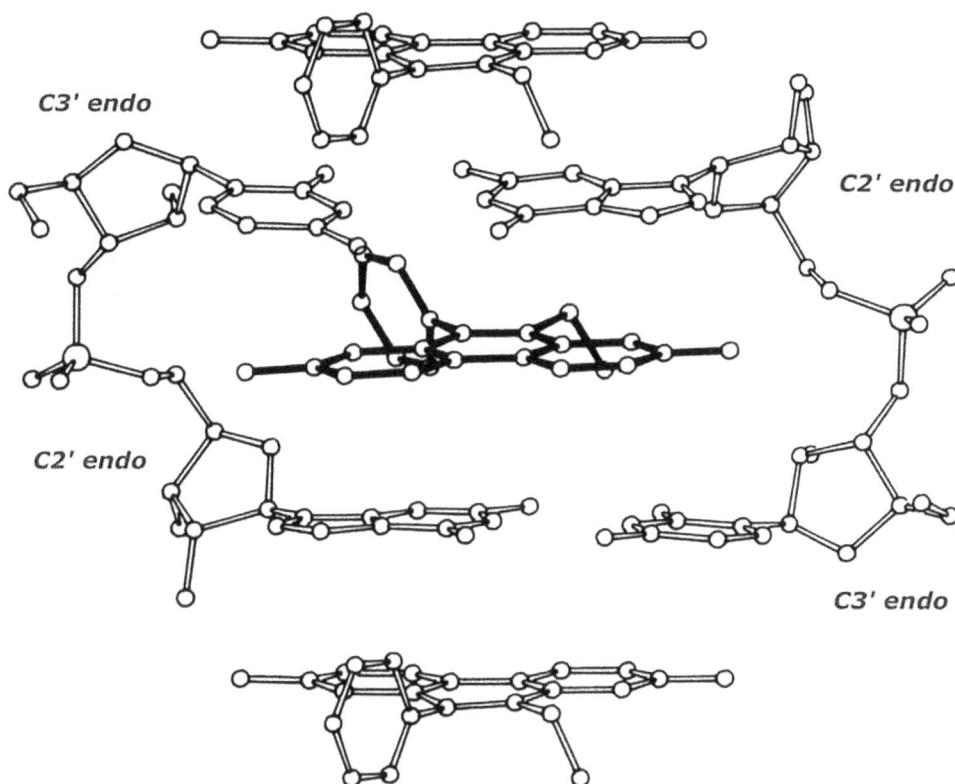

Figure 18: Structure of a 2:2 ethidium-CpG crystalline complex. Cytidylyl deoxyribose residues are *C3' endo* and in the low *anti* conformation, while guanosine deoxyribose residues are *C2' endo* and in the high *anti* conformation. Base-pairs above and below the intercalator are twisted (using as a reference the angle relating interglycosidic (C1') vectors projected on the plane of the ethidium-chromophore) by about 10 degrees and are separated 6.7 Angstroms. The planes of the G-C base pairs above and below the intercalator are not exactly parallel. They are inclined by about 10 degrees, opening into the narrow groove direction. This reflects the steric presence of the phenyl- and ethyl- groups on the intercalated ethidium molecule.

The invariance of the beta-geometry in this series of studies argues that the beta- structural element is a particularly stable-structure that can accommodate a large variety of heterocyclic ring-systems without significant alterations in its geometry, and that this geometry forms equally well with either DNA- or RNA- like dinucleoside monophosphates. Two key predictions follow if one uses this information to understand how simple intercalators bind to DNA. The *first* is that simple intercalators unwind DNA at the immediate intercalation site by -26 degrees (104). The *second* is that simple intercalators give rise to neighbor-exclusion binding, its presence most easily detected at saturating drug/DNA binding-ratios [17] (105). See Figure19.

Proflavine clearly behaves differently from acridine-orange when intercalating into iodoCpG, and we have followed this up with additional crystallographic studies with N, N- dimethylproflavine complexed to iodoCpG and d-CpG, as described below (94). It is noted that proflavine intercalates *symmetrically* in our studies, forming hydrogen-bonds to phosphate-oxygens on opposite-chains (the distance observed between phosphate- oxygens is 15.3 Angstroms). This gives rise to a unique intercalative-geometry; one having all ribose sugars in the C3' *endo* conformation (i.e., C3' *endo* (3'-5') C2' *endo*). Base-pairs above and below the intercalator are separated 6.8 Angstroms -- however, they remain

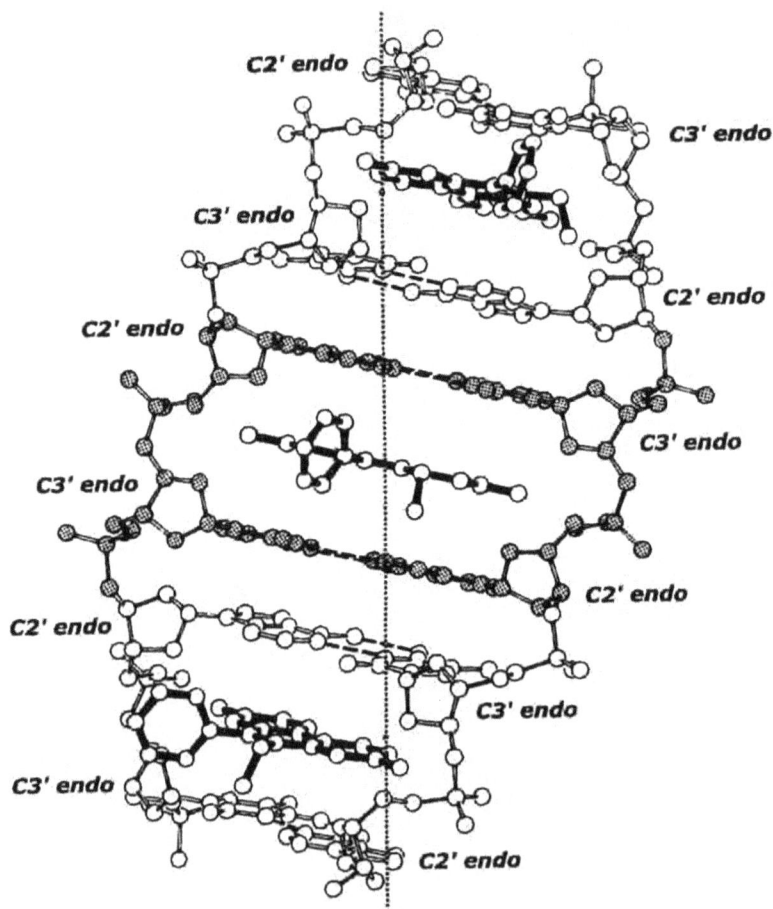

Figure 19: Ethidium-DNA neighbor-exclusion binding-model. This structure has been calculated using the idealized information obtained from X-ray crystallographic studies of the ethidium: dinucleoside- monophosphate complexes (stippled), in combination with the technique of linked-atom least-squares (46). The beta-structural element *plus* ethidium form the asymmetric-unit of the helix – a repeated twist of 47.2 degrees, and a translation of 9.8 Angstroms of this asymmetric-unit along the helix-axis generates the helical- complex shown.

It should be noted that intercalation occurs between every-other base-pair, since binding is restricted to neighboring beta-structural elements. This feature explains the magnitude of DNA stretching and unwinding accompanying neighbor-exclusion binding. Notice that the stereochemistry connecting neighboring beta- structural elements is different (i.e., *C2' endo* (3'- 5') *C3' endo*) – there is no significant stretching or unwinding in this region.

Since DNA in this helical complex contains the beta-structural element as its underlying asymmetric-unit, we have called this structure beta-DNA. As explained earlier, beta-DNA is a metastable and hyperflexible DNA-form that exists in a variety of energy-states. Its highest energy-state (being found in the ethidium-DNA complex above) must be "pinned" by ethidium, in order to be visualized by X-ray crystallography[17].

We have also investigated a similar model that explains double-stranded RNA neighbor exclusion binding. Pointedly, poly d-(A-T) has a B-like DNA structure, whereas poly r-(A-U) has an A-like RNA structure – yet, at high drug/DNA binding ratios – *both* form complexes with ethidium that contain *exactly* one ethidium complexed to every *two* base-pairs. Neighbor- exclusion, therefore, takes place into both synthetic DNA- and RNA-like polymers, again strongly suggesting the presence of an intermediary beta-DNA (or -RNA)form.

twisted 36 degrees. *The angular unwinding component present in the beta- structural element is absent from this conformation. It seems clear that the presence of hydrogen-bonds provides additional energy, which stabilizes this unusual intercalative geometry.* See Figure 20.

Acridine-orange, however, is noted to bind *asymmetrically* to the beta-structural element in these model studies. Being a methylated proflavine-derivative, acridine- orange is unable to hydrogen-bond to phosphate-oxygen atoms on opposite-chains. Instead, acridine-orange intercalates into the beta-structural element, forming tight stereo-specific stacking interactions with adjacent guanine rings. This asymmetry is a feature in common with 9-aminoacridine in our crystallographic studies.

It is possible for proflavine to intercalate into the beta-structural element; however, it is unable to form hydrogen-bonds simultaneously with both phosphate oxygens on opposite-chains (the distance between phosphate-oxygen atoms is 17.3 Angstroms). Such a complex is predicted to be *asymmetric* therefore, proflavine hydrogen-bonding to only one of two phosphate-oxygens. To investigate this, we subsequently synthesized N, N- dimethylproflavine, and examined three different crystalline complexes containing RNA- like and DNA-like self-complementary dinucleoside-monophosphates (98, 99).

We have found N, N dimethylproflavine to intercalate *asymmetrically* into the beta- structural element in all *three* cases. As expected, N, N-dimethylproflavine intercalates asymmetrically into the beta-structural element; however, unexpectedly, its unmethyl- ated amino-group does *not* hydrogen-bond to a neighboring phosphate-oxygen atom (distance, 3.5 Angstroms). Instead, this amino-group hydrogen bonds to a water- molecule. This reflects the absence of coplanarity of both phosphate-oxygen atoms with the N, N- dimethylproflavine molecular least-squares plane. In this structure (and in the other two structures studied), phosphate-oxygen atoms have been found to lie 0.7 Angstroms above (or below) the plane of the N,N-dimethylproflavine-molecule.

Although proflavine is a complex intercalator in these model studies, it acts as a simple intercalator when binding to high molecular weight DNA. This is known since proflavine has been observed to unwind superhelical DNA, in much the same manner as ethidium and acridine orange (56) [18].

B. X-Ray Fiber Diffraction Evidence for Neighbor-Exclusion Binding of a Platinum Metallo-intercalator Reagent to DNA

*Bond, PJ, R Langridge, KW Jennette and SJ Lippard. Natl Acad Sci USA 1975; **72**: 4825- 4829.*

A description of this study has been included in the main text (see IMPLICATIONS IN MOLECULAR BIOLOGY section), and for this reason will not be discussed here.

C. Lengthening and Unwinding of Duplex DNA in Complexes with recA Protein

*Stasiak, A., DiCapua, E. and Koller, T., Institute for Cell Biology, Swiss Federal Institute of Technology, Hoenggerberg, CH-8093 Zurich, Switzerland Cold Spring Harbor Symp Quant Biol 1983; **47**: 811- 820.*

PROFLAVINE　　　**ACRIDINE ORANGE**

(a)　　　(b)

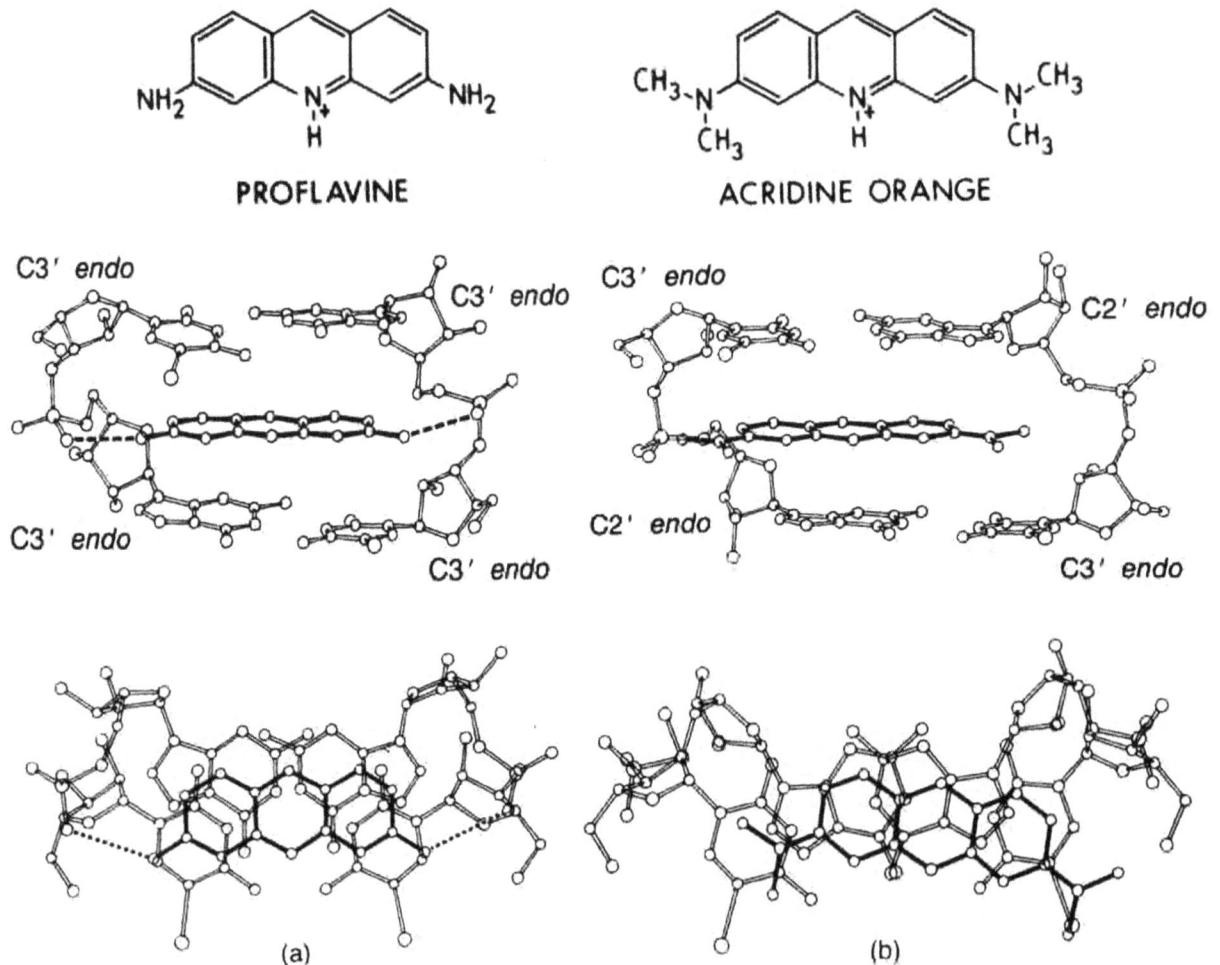

Figure 20: A comparison between the structural-information obtained for proflavine and acridine-orange complexed to the dinucleoside-monophosphate, iodo-CpG. Proflavine intercalates *symmetrically* into a distorted A-type structure, forming hydrogen-bonds to phosphate-oxygens on opposite chains (lower-left and upper-left); whereas, acridine- orange, being a methylated proflavine-derivative, intercalates a*symmetrically* into the beta- structural element (lower-right and upper-right). We have proposed the presence of hydrogen-bonds in the proflavine-complex to pull oppositely oriented sugar-phosphate chains closer-together, this giving rise to the unique intercalative-geometry that is observed. See text for further discussion.

This elegant electron-microscopic study of the *recA* protein of *Escherichia coli* complexed to circular double-stranded nicked- relaxed DNA and to negatively- superhelical DNA- molecules establishes the existence of a DNA-form clearly different from either A- or B- DNA (43).

RecA protein, in the presence of the non-hydrolyzable ATP analog, adenosine 5' gamma- thio triphosphate, binds to nicked relaxed circular-DNA with high-cooperativity, forming an organized helical-structure in which double-stranded DNA is stretched one and one- half times (see Figure 20 a). The complex can be visualized with platinum- shadowed specimens (or with negative- staining using phosphotungstic-acid) and appears as a cross-striated (i.e., as a zig-zag) protein- DNA complex.

N, N-DIMETHYLPROFLAVINE

(a)

(b)

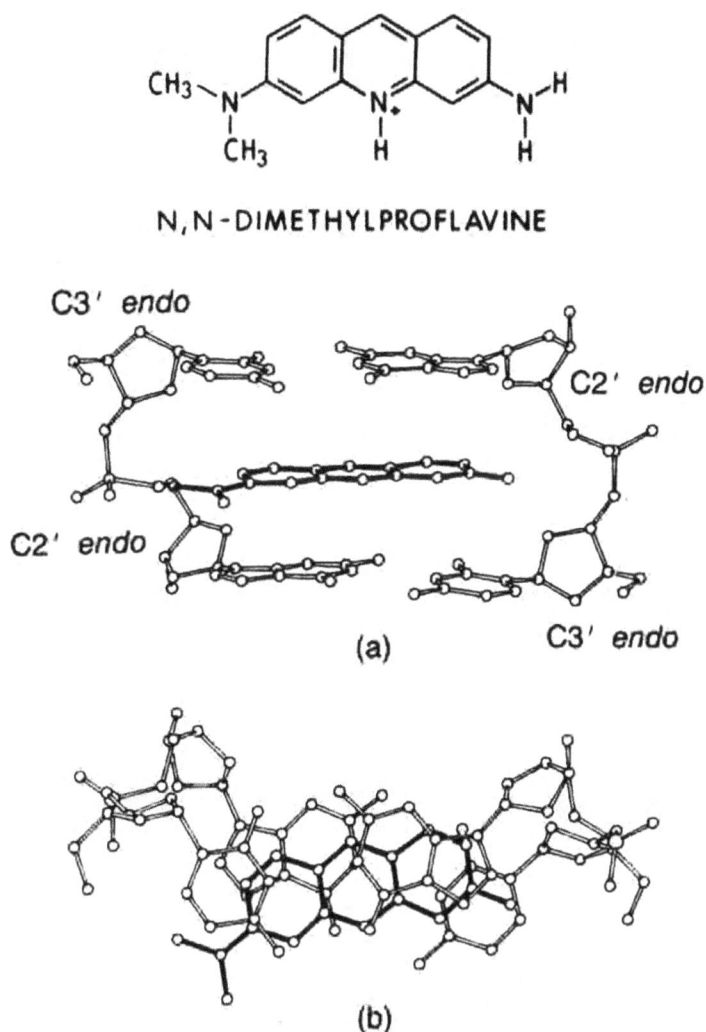

Figure 21: Structural information from the N, N-dimethylproflavine-d-CpG crystalline complex.

(a) Skew view to base pairs and drug molecule, showing the beta-structural element.

(b)perpendicular-view to base-pairs and drug-molecule, showing N, N-dimethylproflavine stacked between the base-pairs within the beta-structural element. This figure should be compared with the information shown in Figure 20. Similar information has been obtained from crystallographic studies of complexes containing N, N-dimethylproflavine and iodoCpG and d-CpG (98, 99). See text for additional discussion.

[18] We are aware of the controversy in the literature regarding the role sugar-puckering plays in determining the intercalation stereochemistry of drugs and dyes into DNA (106-108). Nevertheless, the evidence presented here clearly supports our assertion that simple- intercalators (with the exception of proflavine) bind to the beta-structural element in all the model dinucleoside monophosphate studies done by our research group and the MIT group to date.

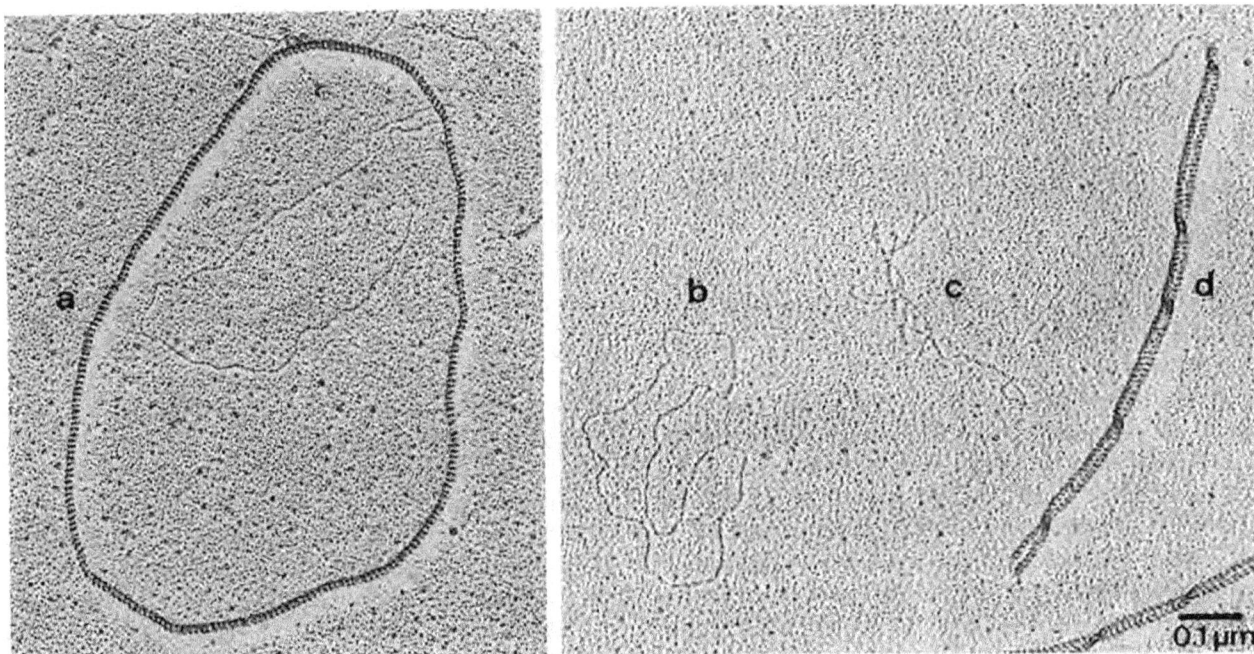

Figure 22: Survey electron micrographs of a specimen of *recA* protein-DNA complexes formed with relaxed circular and covalently closed circular DNA at a protein: DNA ratio (w/w) of 25:1. Two types of circular complexes are observed, the open circles (a) and the folded circles (d), with 16-18 cross striations per turning-over point. Note also twisted (c) and relaxed (b) free DNA molecules. Preparation of the platinum-shadowed specimen was carried out as described by DiCapua et al., 1982 (reproduced, with permission, from Stasiak, A., et al, 1983 (43).

There are 267 cross-striations on the plasmid, which contains 4,961 base pairs. This means that each striation (estimated to contain about 6.2 *recA* monomers) covers 18.6 base pairs, and that the helical structure (inferred from examination of the symmetrical diffraction patterns obtained from scanning-transmission electron microscopic-images negatively- stained segments of the complex) contains 18.6 base pairs per turn in approximately 99 Angstroms.

This helix repeat agrees very closely with the value of 95 Angstroms, obtained by dividing the average contour-length of each circular-complex (25,530 Angstroms) by the number of cross striations (267).

The binding of *recA* to double-stranded DNA also unwinds superhelical-DNA. This is shown in Figure 23d, which shows a negatively-superhelical DNA-molecule bound by the *recA* protein to form a gentle left-handed superhelical-structure, until further binding is inhibited by the appearance (and the energetics) of a left-handed positively- superhelical DNA-tail.

One can quantitate this measurement by studying the binding of *recA* to negatively- superhelical DNA-molecules with known linking-numbers. This is achieved by using the nicking-closing extract from rat-liver on nicked-relaxed circular-DNA molecules, in the presence of increasing concentrations of ethidium-bromide. After purification, it is possible to separate topoisomers having different linking-numbers by gel electrophoresis and to analyze their binding to *recA*. See Figures 22 and 23.

Topoisomers having smaller linking-numbers can also be estimated by studying the percent covering of DNA molecules by *recA* as a function of the ethidium bromide concentration in the

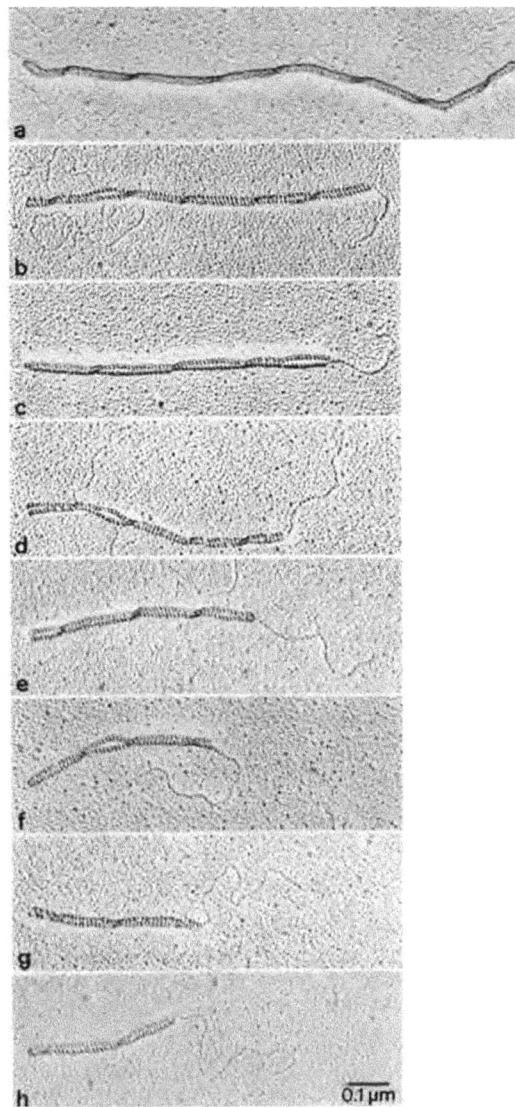

Figure 23: Complexes of *recA* protein and covalently closed circular DNA (pBR G) of various linking numbers (a-h). Complexes with covalently closed circular DNA obtained by relaxation with nicking-closing extract from rat liver (Champoux and McConaughy, 1976) in the presence of increasing concentrations of ethidium bromide. (a) Folded circle obtained with nicked DNA. Note that in b-e, the rod like striated structures arise from the side-by-side aggregation of neighboring segments of the same circular *recA* protein-DNA complex, whereas the thin filaments represent the highly positively supercoiled segments not covered by *recA* protein. The linking numbers of the covalently closed circular DNA samples were 472 (h), 458 (g), 445 (f), and 432 (e), as determined from band-counting gels (Keller, 1975). Samples in this range were used for the plot in Fig. 5. b-d are from covalently closed circular DNA of a higher linking number, which cannot be determined by classic means, but which we have determined as described in Fig. 7. The *recA* protein-DNA complexes were obtained as described in Experimental Procedures [reproduced, with permission, from Stasiak, A., et al, 1983(43)].

nicking-closing reaction in the presence of known amounts of circular-DNA. One can readily see a that there is a direct correlation between the percent-covering of DNA by *recA* and the linking number (i.e., the smaller the linking number, the greater the percent covering of DNA by *recA*).

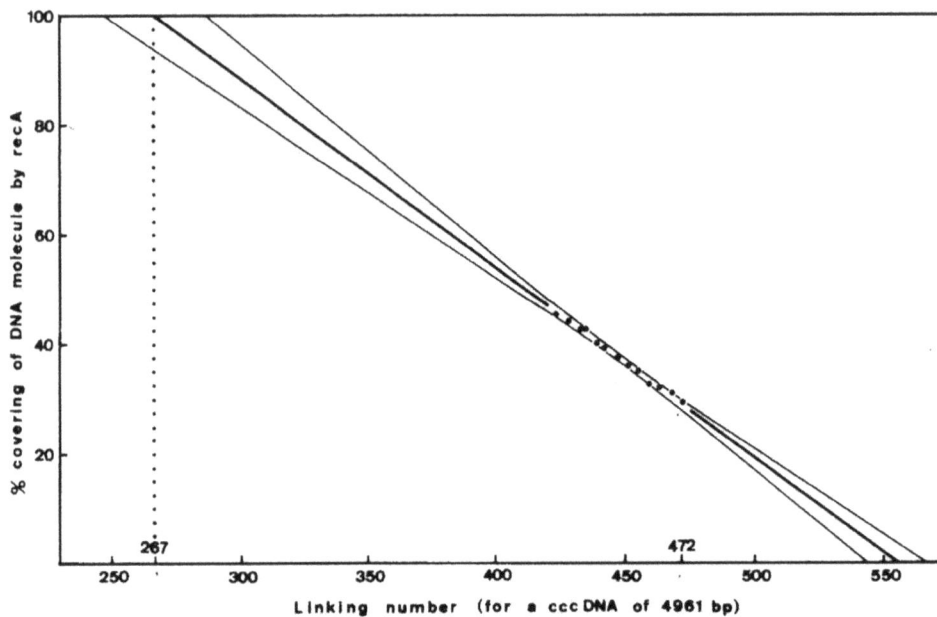

Figure 24: Determination of the linking number of a covalently closed circular DNA molecule fully covered with *recA* protein. The linking numbers of the various covalently closed circular DNAs produced with nicking-closing enzyme and ethidium bromide were calculated by assuming 10.5 bp per turn of B-DNA in solution (Wang, 1979; Rhodes and Klug, 1980), which results in 472 turns for 4961 bp of relaxed DNA. Subtracting from this value the number of superturns determined from the maximum of the distribution on band-counting gels (Keller, 1975) yields the linking number. For each topoisomers family, the extent of covering with *recA* protein was determined from 15-25 molecules. The standard deviation of these measurements was less than 3%. This is also what one would expect from the topoisomers distribution in the families visible on the gels. Extrapolation to 100% covering with *recA* protein leads to a value of 267 +/- 19.2 for the linking number of a totally covered molecule. The confidence interval was drawn assuming 95% confidence (reproduced, with permission, from Stasiak, A., et al, 1983 (43)

Note that at 100% covering the estimated linking-number is 267. One can calculate the unwinding of DNA associated with the binding of *recA* to DNA in such a structure, assuming B-DNA in solution has 10.5 base pairs per turn (and therefore, having a linking- number of 472 for 4,961 base pairs). The unwinding predicted is $[(472-267) \times 360]/ 4,961 = 14.9$ degrees per base-pair, or about 30 degrees every-other base-pair. Their model for the recA DNA complex is shown in Figure 25.

D. *Physical Evidence Concerning the Binding of Irehdiamine and Other Steroidal-Diamines to DNA*

1) Mahler HR, G Green, R Goutarel, Q Khuong-Huu. Nucleic Acid Small Molecule Interactions. VII. Further Characterization of Deoxyribonucleic Acid - Steroidal Diamine Complexes. *Biochemistry* 1968; **7**: 1568-1582.

This paper, along with others in this series of papers, describes early studies done to characterize the nature of a series of steroidal-diamine complexes with DNA. These studies provided evidence for the existence of partial-intercalation by these steroidal- diamines into the lowest-energy beta-DNA form (55).

Dr. Mahler and his associates have shown that:

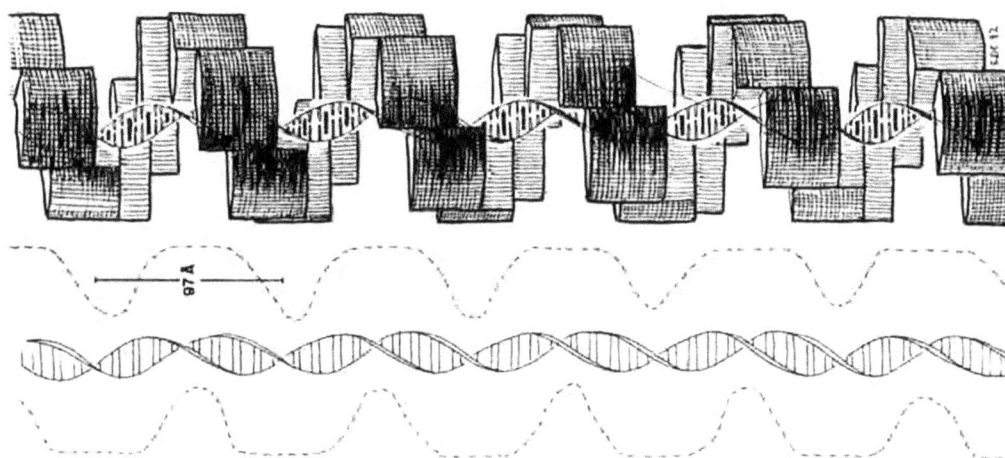

Figure 25: Model of the *recA* protein-DNA helix (reproduced, with permission, from Stasiak, A., et al, 1983 (43)).

irehdiamine and other steroidal-diamines bind tightly to DNA under low-salt conditions, as evidenced by a hyperchromic-effect at 260 millimicrons. The magnitude of this hyperchromic-effect is in the same order as that seen with DNA-melting, suggesting significant base-pair unstacking.

Irehdiamine and other steroidal-diamines form stoichiometric-complexes that contain one steroidal-diamine for every two base-pairs (i.e., one steroidal diamine/ four phosphates). Very similar observations are observed with simple-intercalators, suggesting neighbor-exclusion binding. Another complex has been detected at higher steroidal- diamine - DNA ratios. This complex is less stable to thermal-melting, and, although structurally uncharacterized, could be a superhelical variant of the preceding form.

The melting temperature of the first steroidal-diamine - DNA complex is about 20 degrees higher than native DNA, suggesting that complex-formation may interfere with the nucleation and propagation of DNA-melting. Irehdiamine and other steroidal- diamines show little effect on DNA-viscosity. As a function of steroidal-diamine concentration, the intrinsic-viscosity first falls (about 10%), and then rises to the initial intrinsic-viscosity for native-DNA. Very different observations are observed with simple-intercalators (a large rise in viscosity accompanies binding), suggesting that complex formation neither stiffens nor lengthens DNA.

Spectroscopic (ORD) measurements suggest DNA to have a very different conformation in these complexes, when compared with either A- or B- DNA.

And, finally, irehdiamine reverses proflavine-induced mutations in T2 bacteriophage, suggesting that irehdiamine binds to DNA in a manner similar to intercalation.

2) MJ Waring. Variation of Supercoils in Closed Circular DNA by Binding of Antibiotics and Drugs: Evidence for Molecular Models Involving Intercalation. J Mol Biol 1970; **54**: 247-279.

This study (of the unwinding of negatively-superhelical DNA by intercalating-agents) is the most complete of its kind. Irehdiamine is observed to unwind DNA, much like other

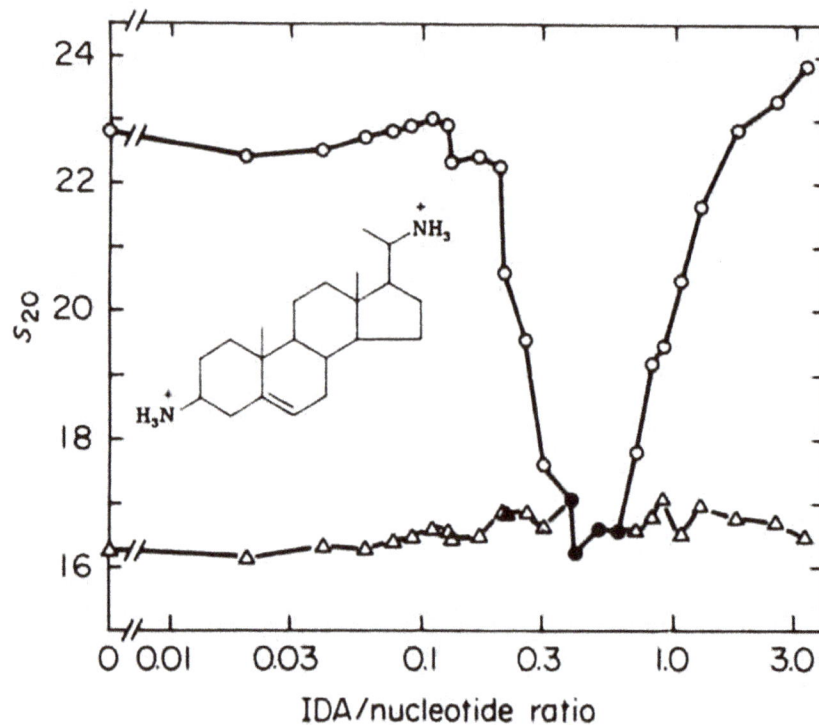

Figure 26: Sedimentation behavior of superhelical DNA in the presence of increasing concentrations of irehdiamine A. Circles indicate closed circular PM2 sedimentation coefficients that are negatively superhelical; triangles indicate closed circular PM2 DNA sedimentation coefficients when DNA is nicked relaxed (reproduced, with permission, from Waring, M.J., 1970 (56)).

planar-intercalators, unwinding DNA roughly half as much as ethidium (see Figure 26).

This suggests that irehdiamine binds to DNA by *partial-intercalation* into the beta- structural element. Our model predicts an angular-unwinding of about 12 degrees – in good agreement with the data.

10. Addendum

Premeltons are examples of "kink-antikink bound-states" (i.e., or equivalently, "breather solitons") in DNA, these arising spontaneously within the early melting regions of DNA to nucleate site- specific DNA-melting. Their presence allows one to understand how drugs and dyes intercalate into DNA, and also to understand how nucleases such as the micrococcal- nuclease and the pancreatic DNase 1 enzymes recognize and cleave early melting regions in naked DNA molecules.

Energies necessary for the formation of the premelton in DNA molecules come from Brownian motion, excited by solvent collisions at normal (i.e., kT) energies.

We envision DNA in solution to be continually bombarded by solvent collisions along its length. Although at first glance one might expect the average excitation force to be zero, this is not the case in the microscopic domain, where DNA is continually experiencing unbalanced forces (i.e., the Brownian force).

Since the collision cross-sectional area of DNA is small (i.e., the diameter being about 20 Angstroms), relatively few solvent molecules impinge on its surface at short time intervals. Moreover, the flexibility of DNA is highly anisotropic. Due to this anisotropy, most solvent collisions are expected to have little effect on DNA structure, exciting only small amplitude normal mode motions in functional groups. These are expected to damp readily through solvent interactions. There are, however, small windows of collisions that deform DNA nonlinearly. We believe these collisions to hit DNA from both wide and narrow groove directions, striking DNA along dyad axes located between adjacent base pairs.

Such collisions give rise to nonlinear pulses in DNA (also called solitons, or solitary excitations), which move along the polymer chain with a velocity significantly less than the speed of sound.

These contain a modulated beta-alternation in sugar puckering along both polynucleotide chains, and are nontopological – that is, although these excitations unwind DNA, this is counterbalanced by right-handed superhelical writhing to keep the linking invariant. Energies stored and transmitted by such intrinsic locally coherent excitations can travel considerable distances along DNA with minimal dissipative loss, since they are largely internal to the polymer structure. In addition, such nonlinear pulses remain "robust", since the nonlinearity present in the sugar-pucker conformations acts to minimize dispersion- effects.

The shape of the energy density profile accompanying low energy solitary excitations is expected to be sensitive to the nucleotide base sequence in DNA. This is because different DNA regions contain different base-stacking energies, and have, therefore, different intrinsic flexibilities. Energy-density profiles of traveling solitary excitations are expected to sharpen up (i.e., the leading edge of the excitation traveling more slowly than the trailing edge) within regions that start out by being more flexible than other regions. This acts to deform DNA structure, and to enhance the lifetime of these excitations.

This increases the probability that, with increasing temperature, still larger excitations can form from the coalescence of additional solitary excitations that arrive in these regions. The appearance of these larger excitations deforms DNA regions still further and gives rise to even greater flexibility in their most central regions. Eventually, with increasing temperature, premeltons arise that contain hyperflexible (liquid-like) beta-DNA cores, surrounded by phase-boundaries termed "kink" and "antikink".

Premeltons have longer lifetimes, and, with increasing temperature, become nucleation- centers for DNA-melting. Traveling solitary excitations, originating from solvent collisions at earlier times and from more remote locations, enter the premelton, their energies being trapped within the core. This energy continues to enlarge beta-DNA core-regions and increases the amplitude of breather-motions, leading to DNA "breathing". Finally, with increasing temperature, permanently melted single-stranded DNA regions appear within the premelton. Their appearance signals the onset of DNA melting. We have called these larger structural solitons, "meltons".

With increasing temperature; meltons continue to trap energy from entering solitary excitations that have arisen elsewhere (and at earlier times) along B-DNA. Their energies are used to separate kink from antikink while lengthening internally melted single-stranded DNA- regions. Finally, with increasing temperature, meltons coalesce and all DNA becomes single- stranded. Single-stranded DNA is extremely flexible (i.e., entropic), and can be likened to a gas- like phase.

In summary then, my thermal mechanism predicts the existence of three structural phases for DNA: the Watson-Crick A- and B- forms (solid), the hyperflexible beta-DNA form (liquid), and the highly entropic single-stranded melted-DNA form (gas). The beta-DNA phase within the premelton is predicted to play a key role in nucleating DNA-melting.

11. References

1) Davydov AS. The theory of contraction of proteins under their excitation. J Theor Biol. 1973; **38**: 559-569.

2) Davydov AS. Solitons in Molecular Systems. Physica Scripta. 1979; **20**: 387-394.

3) Davydov AS. Solitons in Molecular Systems, 2nd Edition 1991; Kluwer Academic Publishers Reidel, Dordrecht.

4) Englander, SW, NR Kallenbach, AJ Heeger, JA, Krumhansl and S Litwin. Nature of the open state in long polynucleotide double helices: possibility of soliton excitations. *Proc Natl Acad Sci USA*. 1980; 77: 7222-7226.

5) Krumhansl JA and D Alexander, in Structureand Dynamics: Nucleic Acids and *Proteins, Adenine Press, Inc.,* pp. 61-80 (1983).

6) Takeno S and S Homma. Topological Solitons and Modulated Structure of Bases in DNA double Helices: A Dynamic Plane Base-Rotator Model. Prog Theor Phys. 1983; 70: 308-311.

7) Yomosa S. Soliton Excitations in Deoxyribonucleic Acid Phys Rev A. 1983; **27**: 2120-2125.

8) Yomosa, S. Soliton excitations in deoxyribonucleic acid (DNA) Phys Rev A. 1984; **30**: 474-480.

9) Khan A, D Bhaumik and Dutta-Roy. The possible role of solitonic processes during A to B conformational changes in DNA. Bull Math Biol. 1985; **47**: 783-789.

10) Zhang CT. Solitary Excitations in Deoxyribonucleic Acid (DNA) Double Helices. Phys Rev A 1987; **35**, 886-891.

11) Muto V., Holding, K., Christiansen, P., and Scott, A. Solitons in DNA. *JBiomol Structure & Dynamics* 1988; **5**: 873-894 (1988).

12) Muto V, Scott, A.C, and P Christiansen. Thermally generated solitons for a toda lattice model of DNA. *Phys Rev Lett A.* 1989; **136**: 33-36.

13) Peyrard M and AR Bishop. Statistical Mechanics of a nonlinear model forDNA denaturation A. *Phys Rev Lett* 1989; **62**: 2755-2758.

14) Muto V, Lomdahl, P, and P Christiansen. Two-dimensional discrete model for DNA dynamics: longitudinal wave propagation and denaturation. *Phys Rev A* 1990; **42**: 7452-7458.

15) Salerno M. Discrete model for Promoter-DNA dynamics. *Phys Rev A* 1991; **44**:*5292*-5297.

16) Salerno M. Dynamic properties of DNA promoters. *Phys Rev Lett A* 1992; **167:** 49-53.

17) Dauxois T, Peyrard, M, and Bishop, AR. Dynamics and thermodynamics of a nonlinear model for DNA denaturation. *Phys Rev E* 1993; 684-695.

18) Kalosakas G, Rasmussen, KO, Bishop, AR, Choi, CH, and A Usheva. Sequence specific thermal fluctuations identify start sites for DNA transcription. *Europhys Lett* 2004; **68:** 127-133.

19) Yakushevich LV. *Nonlinear Physics of DNA, Second Edition, Chichester and New York: Wiley* 2005.

20) Pang X-F and Y-P Feng. *J Biomol Struc & Dyn.* Solitons in DNA. 2007; **25:** 435- 451.

21) Scott AC, FYF Chu and DW McLaughlin. The soliton: a new concept in applied science. *Proc IEEE* 1973; **61:** 1443-1483.

22) Barone A, F Esposito, CJ Magee and AC Scott. Theory and application of the Sine- Gordon equation. *Riv Nuovo Cimento* 1973; **1:** 227-267.

23) Scott AC. *Physics Reports. Theory and application of the sine-Gordon equation* 1992; **217:** 1-67 (1992).

24) Scott, AC. *Nonlinear Science: Emergence and Dynamics of Coherent Structures, Second Edition, Oxford University Press, Oxford, New York* 2003.

25) Scott AC. *Encyclopedia of Nonlinear Science,* ed. A. C. Scott Routledge Taylor & Francis Group, New York (2005).

26) Scott AC. *The Nonlinear Universe: Chaos, Emergence, Life,Springer-Verlag, Berlin, Heidelberg, New York* (2007).

27) Sobell HM, ED Lozansky and M Lessen. Structural and energetic considerations of wave propagation in DNA. *Cold Spring Harbor Symp Quant Biol* 1979; **43:** 11-19.

28) Sobell HM, TD Sakore, SC Jain, A Banerjee, KK Bhandary, BS Reddy, and E Lozansky. β-Kinked DNA—a structure that gives rise to drug intercalation and DNA breathing—and its wider significance in determining the premelting and melting behavior of DNA. *Cold Spring Harbor Symp Quant Biol* 1983; **47:** 293- 314.

29) Banerjee A and HM Sobell. Presence of nonlinear excitations in DNA structure and their relationship to DNA melting and to drug intercalation. *J Biomol Struc & Dyn* 1983; **1:** 253-262.

30) Sobell HM. Actinomycin and DNA transcription. *Proc Natl Acad Sci USA* 1985; **82:** 5328-5331.

31) Sobell, HM. *Biological Macromolecules & Assemblies Volume 2: Nucleic Acids & Interactive Proteins, ed. A.A. Jurnak and A. McPherson, John Wiley & Sons pp.* 172-232 1985.

32) Sobell, HM. DNA premelting, in *Encyclopedia of Nonlinear Science ed. A.C. Scott Routledge* 2005; 225- 227.

33) Lerman LS. Structural considerations in the interaction of DNA and acridines. *J Mol Bio 1961*; **13**: 18-30.

34) Printz MP, PH von Hippel. Hydrogen exchange studies of DNA structure. *Proc Natl Acad Sci USA* 1965; 53: 363-367.

35) Fiel RJ, Munson, BR. Binding of meso-tetra (4-N-methylpyridyl) porphine to DNA. *Nucleic Acids Res* 1980; 8: 2835-2842.

36) Waring MJ, LPG Wakelin. Echinomycin: a bifunctional intercalating antibiotic. *Nature* 1974; 252: 653-657.

37) White JH. Self-Linking and the Gauss Integral in Higher Dimensions. *Am J Math* 1969; **91**: 693-728

38) Fuller FB. The writhing of a space curve. *Proc Natl Acad Sci USA* 1971; **68:** 815- 818.

39) Crick FHC. Linking numbers and nucleosomes. *Proc Natl Acad Sci USA* 1976; **73:** 2639-2643.

40) Frank-Kamenetskii MD, A Vologodskii. *Sov Phys Usp* 1981; **24**: 679-69.

41) Su WP, JR Schrieffer and AJ Heeger. Soliton excitations in polyacetylene. *Phys Rev B* 1980; **22**: 2099-2111.

42) Bishop AR, *in Nonlinear Electrodynamics in Biological Systems Ed. W. R. Adey and F. Lawrence, Plenum Press pp.* 155-175 (1984).

43) Stasiak A, E DiCapua and T Koller. Unwinding of duplex DNA in complexes with recA protein. *Cold Spring Harbor Symp Quant Biol* 1983; **47,** 811- 820.

44) Kiianitsa K and A Stasiak. Helical repeat of DNA in the region of homologous pairing. *Proc Natl Acad Sci USA* 1997; **94:** 7837-7840.

45) Radding CM, J Flory, Wu, A, Kahn, R, DasGupta, C, Gonda, D, Bianchi, M and SS Tsang. *Cold Spring Harbor Symp Quant Biol* 1983; **42**: 821-828.

46) Smith PJC, S Arnott. LALS: A linked-atom least squares reciprocal space refinement incorporating stereochemical restraints to supplement sparse diffraction data. *Acta Cryst A* 1978; **34**: 3-10.

47) Fogel MB, Trullinger SE, Bishop AR, and Krumhansl JA. Classical Particle like Behavior of Sine-Gordon Solitons in Scattering Potentials and Applied Fields. *Phys Rev Lett* 1976; 3**6:** 1411-1414.

48) Platt JR. Possible Separation of Intertwined Nucleic Acid Chains by Transfer-Twist. *Proc Natl Acad Sci* 1955; **41:**181-185.

49) Lilley DMJ. Dynamic, sequence-dependent DNA structure as exemplified by cruciform extrusion from inverted repeat in negatively supercoiled DNA. *Cold Spring Harbor Symposium Quant Biol* 1982; 47: 101-112.

50) Hentschel CC. Homo-copolymer sequences in the spacer of a sea urchin histone gene repeat is sensitive to S1 nuclease. *Nature* 1982; **295:** 714-716.

51) Wang A, GJ Quigley, GJ, Kolpak, FJ, Crawford, JL, van Boom, JH, van der Marel, G, Rich,

A. Molecular structure of a left-handed double helical fragment at atomic resolution. *Nature* 1979; **282:** 680-686.

52) Wells RD, R Brennan, KA Chapman, TC Goodman, PA Hart, W Hillen, DR Kellogg, MW Kilpatrick, RD Klein, J Klysik, PF Lambert, JE Larson, JJ Miglietta, SK Neuendorf, TR O'Connor, CK Singleton, SM Stirdivant, CM Veneziale, RM Wartell, W Zacharias. Left- handed DNA helices, supercoiling, and the B-Z junction. *Cold Spring Harbor Symp* Quant *Biol* 1983; **47,** 77-84.

53) Pohl F, TM Jovin, W Baehr, JJ Holbrook. Ethidium bromide as a cooperative effector of a DNA structure. *Proc Nat Acad Sci USA* 1972; **69:** 3805-3809.

54) Pohl FM, TM Jovin. Salt-induced co-operative conformational change of a synthetic DNA: Equilibrium and kinetic studies with poly(dG-dC). *J Mol Biol* 1972; 67: 375- 396.

55) Mahler HR, G Green, R Goutarel, Q Khuong-Huu. Nucleic acid- small molecule interactions. VII. Further characterization of deoxyribonucleic acid-diamino steroid complexes. *Biochemistry* 1968; **7:** 1568- 1582.

56) Waring MJ. Variation of the supercoils in closed circular DNA by binding of antibiotics and drugs: Evidence for molecular models involving intercalation. J Mol Biol 1971; **54:** 247-279.

57) Waring, MJ. Echinomycin: A bifunctional intercalating antibiotic. *J Mol Biol* 1970; **54,** 247-279.

58) Waring, MJ, JW Chisholm. Uncoiling of bacteriophage PM2 DNA by binding of steroidal diamines. *Biochim Biophys Acta* 1972; **262:** 18-23.

59) Waring, MJ, Henley, SM. Stereochemical aspects of the interaction between steroidal diamines and DNA. *Nucleic Acids Res* 1975; **2:** 567-586.

60) Dattagupta, N, M Hogan, DM Crothers. Does irehdiamine kink DNA? *Proc Natl Acad Sci* 1978; *75:* 4286-4290.

61) Patel, DJ, LL Canuel. Steroid diamine-nucleic acid interactions: partial insertion of dipyrandium between unstacked base pairs of the poly(dA-dT) duplex in solution. *Proc Natl Acad Sci USA* 1979; **76,** 24-28.

62) Kypr J, M Vorlickova. Conformations of alternating purine-pyrimidine DNAs in high- CsF solutions and their reversal by dipyrandium, ethidium and high temperature *Biochim et Biophys Acta* 1985; **838:** 244-251.

63) Sigman DS, CHB Chen. Chemical nucleases: New reagents in molecular biology. *Ann Rev Biochem* 1990; 59: 207-236.

64) Jessee B, G Gargiulo, F Razvi and A Worcel. Analogous cleavage of DNA by micrococcal nuclease and a 1-10-phenanthroline-cuprous complex. *Nucleic Acids Res* 1982; **10**: 5823-5834.

65) Glikin GC, G Gargiulo, L Rena-Descalzi, A Worcel. Escherichia coli single-strand binding protein stabilizes specific denatured sites in superhelical DNA. *Nature London* 1983; 303: 770-774.

66) Samai B, A Worcel, C Louis, P Schedl. Chromatin structure of the histone genesin **D.** melanogaster. *Cell* 1981; **23:** 401-409.

67) Cartwright IL, Elgin, S.C.R. Analysis of chromatin structure and DNA sequence organization: use of the 1,10-phenanthroline-cuprous complex. *Nucleic Acids Res* 1982; **10**, 5835-5852. Analysis of chromatin structure and DNA sequence organization: use of the 1,10-phenanthroline-cuprous complex.

68) Elgin SCR, Cartwright, G, Fleischmann, K Lowenhaupt, MA Keene. Cleavage reagents as probes of DNA sequence organization and chromatin structure: Drosophila melano-gaster locus 67B1. *Cold Spring Harbor Symp Quant Biol* 1983; **47**: 529-538.

69) Miller OL, Jr, BR Beatty. Visualization of nucleolar genes. *Science* 1969; **164:** 955-957.

70) Liu LF, JC Wang. Supercoiling of the DNA template during transcription. *Proc Natl Acad Sci USA* 1987; **84**:7024-7027.

71) Weintraub H, M Groudine. Chromosomal subunits in active genes have an altered conformation. *Science* 1976; **193:** 848-853.

72) Bond PJ, R Langridge, KW Jennette, SJ Lippard. X-ray fiber diffraction evidence for neighbor-exclusion binding of a platinum metallointercalator agent to DNA. *Proc Natl Acad Sci USA* 1975; 72: 4825-4829.

73) Sobell, HM, SC Jain, TD Sakore, and CE Nordman. Stereochemistry of actinomycin–DNA Binding. *Nature New Biol* 1971; **205:** *231*-200.

74) Jain, SC, HM Sobell. Stereochemistry of actinomycin binding to DNA. I. Refinement and further structural details of the actinomycin-deoxyguanosine crystalline complex. *J Mol Biol* 1972; 68: 1-20.

75) Sobell HM, SC Jain. Stereochemistry of actinomycin binding to DNA. II. Detailed molecular model of actinomycin-DNA complex and its implications. J Mol Biol 1972; **68:** 21-34.

76) Sobell HM. How actinomycin binds to DNA. *Scientific American* 1974; **231**: 82-91.

77) Wells, RD and JE Larson. Studies on the binding of actinomycin D to DNA and DNA model polymers. *J Mol Biol* 1970; **49**:319-342.

78) Kamitori, S and FJ Takusagawa. Multiple binding modes of anticancer drug actinomycin D: x-ray, molecular modeling, and spectroscopic studies of d(GAAGCTTC)2 --actinomycin D complexes and Its host DNA. *Am Chem Soc.* 1994; **116:** 4154-4165.

79) Shinomiya, M, Chu, W, Carlson, RG, Weaver, RF and Takusagawa, F. Structural, physical, and biological characteristics of RNA.DNA binding agent N8-actinomycin D. *Biochemistry* 1995; **34**, 8481-8491.

80) Takusagawa, F, KT Takusagawa, RG Carlson and RF Weaver. Selectivity of F8-Actinomycin D for RNA: DNA Hybrids and Its Anti-Leukemia Activity. *Bioorg Med Chem* 1997; **5:** 1197-1207.

81) Hou, M-H, H Robinson, Y-G, Gao and AH Wang. Crystal structure of actinomycin D bound to the CTG triplet repeat sequences linked to neurological diseases. *Nucleic Acid Research* 2002; **30**: 4910-4917.

82) Wang, AH-J, G Ughetto, GJ Quigley, T Hakoshima, GA Van Der Marel, JH Van Boom and A Rich. The molecular structure of a DNA triostin A complex. *Science* 1984; **225:** 1115-1121.

83) Wang, AH, G Ughetto, GJ Quigley and A Rich. Interactions of quinoxaline antibiotic and DNA: the molecular structure of a triostin A-d(GCGTACGC) complex. *J Biomol Struct Dyn* 1986; **4:** 319 -342.

84) Cuesta-Seijo, JA, and GM Sheldrick. Structures of complexes between echinomycin and duplex DNA. *Acta Crystallogr Sect D* 2005; **61**: 442-448.

85) Quigley, GJ, AH Wang, G Ughetto, G van der Marel, MJH van Boom and Rich, Molecular structure of an anticancer drug-DNA complex: daunomycin plus d(CpGpTpApCpG). *Proc Nat Acad Sci USA* 1980; **77**: 7204-7208.

86) Moore, MH, WN Hunter, BL d'Estaintot and O Kennard. DNA-drug interactions. The crystal structure **of** d(CGATCG) complexed with daunomycin. *J Mol Biol* 1989; **206**: 693-705.

87) Frederick CA, LD Williams, G Ughetto, GA, GA van der Marel, JH van Broom, A Rich, AH Wang. Structural comparison of anticancer drug-DNA complexes: adriamycin and daunomycin. *Biochemistry* 1990; 29: 2538- 2549.

88) Tsai C-C, SC Jain, HM Sobell. X-ray crystallographic visualization of drug-nucleic acid intercalative binding: structure of an ethidium-dinucleoside monophosphate crystalline complex, ethidium: 5-iodouridylyl (3'-5') adenosine. *Proc Natl Acad Sci USA* 1975; **72**: 628-632.

89) Tsai C-C, SC Jain, TD Sakore, SC Jain, S.C, C-C Tsai, C.-C, HM Sobell. Mutagen- nucleic acid intercalative binding: structure of a 9-aminoacridine: 5-iodocytidylyl (3'-5') guanosine crystalline complex. *Proc Natl Acad Sci USA* 1984; **74:** 188-192

90) Tsai C-C, SC Jain, HM Sobell. Visualization of drug-nucleic acid interactions at atomic resolution. I. Structure of an ethidium dinucleoside monophosphate crystalline complex, ethidium: 5-iodouridylyl (3'-5') adenosine. *J Mol Biol* 1977; **114**: 301-315.

91) Jain, SC, C-C Tsai, HM Sobell. Visualization of drug-nucleic acid interactions at atomic resolution. II. Structure of an ethidium dinucleoside monophosphate crystalline complex, ethidium: 5-iodocytidylyl (3'–5') guanosine. *J Mol Biol* 1977; **114**: 317-331.

92) Sobell HM, C.-C Tsai, SC Jain, SG Gilbert. Visualization of drug-nucleic acid interactions at atomic resolution. III. Unifying structural concepts in understand drug-DNA interactions and their broader implications in understanding protein-DNA interactions. *J Mol Biol* 1977; **114**: 333-303.

93) Sakore TD, BS Reddy, HM Sobell. Visualization of drug-nucleic acid interactions at atomic resolution. IV. Structure of an aminoacridine-dinucleoside mono- phosphate crystalline complex, 9-aminoacridine: 5-iodocytidylyl (3'-5') guanosine. *J Mol Biol* 1979; **135**: 763-785.

94) Reddy, BS, TP Seshadri, TD Sakore, HM Sobell. Visualization of drug-nucleic acid Interactions at atomic resolution V. Structure of *two* aminoacridine-dinucleoside monophosphate crystalline complexes, proflavine: 5-iodocytidylyl (3'-5') guanosine and acridine orange: 5-iodocytidylyl (3'-5') guanosine. J *Mol Biol* 1979; **135**: 787-812.

95) Jain, SC, KK Bhandary, HM Sobell. Visualization of drug-nucleic acid interactions at atomic resolution. VI. Structure of *two* drug-dinucleoside monophosphate crystalline complexes, ellipticine: 5-iodocytidulyl (3'-5') guanosine and 3,5,6,8- tetramethyl-N- methyl-phenanthrolinium: 5-iodocytidylyl (3'-5') guanosine. *J Mol Biol* 1979; **135: 813-840.**

96) Jain, SC, HM Sobell. Visualization of drug-nucleic Acid interactions at atomic resolution VII. Structure of an ethidium dinucleoside monophosphate crystalline complex, ethidium: uridylyl (3'-5') adenosine J *Biomol Struc & Dyn* 1984; **1**: 1161- 1177.

97) Jain, SC, HM Sobell. Visualization of drug-nucleic acid interactions at atomic resolution. VIII. Structures of *two* ethidium dinucleoside monophosphate crystalline complexes containing ethidium: cytidylyl (3'-5') guanosine. *J Biomol Struc & Dyn*1984; **1:** 1179-1194.

98) Bhandary, KK., TD Sakore, HM Sobell, D King, EJ Gabbay, E. Visualization of drug- nucleic acid interactions at atomic resolution. IX. Structures of *two* N, N dimethyl- proflavine: 5-iodocytid-ylyl (3'-5') guanosine crystalline complexes. *J Biomol Struc & Dyn* 1984; **1:** 1195-1217.

99) Sakore, TD, KK Bhandary, HM Sobell. Visualization of drug-nucleic acid interactions at atomic res- olution. X. Structure of a N, N-dimethylproflavine: deoxycytidylyl (3'-5') deoxyguanosine crystal- line complex. *J Biomol Struc & Dyn* 1984; **1**: 1219-1227.

100) Wang, AH, GJ Quigley, A Rich. Atomic resolution analysis of a 2:1 complex of CpG and acridine orange. *Nucleic Acids Res.* 1979; 6: 3879-3890.

101) Wang, AH, J Nathans, G van der Marel, JH van Boom, A Rich. Molecular structure of a double helical DNA fragment intercalator complex between deoxy CpG and a terpyridine platinum compound**.** *Nature* 1978; **276**, 471- 474.

102) Reinhardt, C. Spectroscopic evidence for sequence preferences in the intercalative binding of ethidium bromide to nucleic acids. *Ph.D. Thesis, 1976. University of Rochester, Department of Chemistry.*

103) Krugh TR, C Reinhardt. Evidence for sequence preferences in intercalative binding of ethidium-bromide to dinucleoside monophosphates *J Mol Biol* 1975; **97**: 133- 162.

104) Wang JC. The degree of unwinding of the DNA helix by ethidium. 1. Titration of twisted PM2 DNA molecules in alkaline cesium chloride density gradients. *J Mol Biol* 1974; **89**: 783-801.

105) Crothers DM. Calculation of binding isotherms for heterogeneous polymers. *Biopolymers* 1968; **6**: 575-583.

106) Berman HM, W Stallings, HL Carrell, JP Glusker, S Neidle, G Taylor, A Achari. Molecular and crystal structure of an intercalation complex: proflavine-cytidyl (3' – 5') guanosine. *Biopolymers* 1979; **18**: 2405-2429.

107) Shieh H-S, HM Berman, M Dabrow, S Neidle. The structure of drug- deoxydinucleoside phosphate complex; generalized conformational behavior of intercalation complexes with RNA and DNA fragments. *Nucleic Acids Res* 1980; **8**: 85-97.

108) Schneider B SL Ginell, HM Berman. Low temperature structures of dCpG- proflavine. Conformational and hydration effects. *Biophys Journal* 1992; **63**: 1572-1578.

109) Sobell HM. Premeltons in DNA. *J Struct Funct Genomics* 2016; **17**: 17-31

110) Sobell HM. How actinomycin binds to DNA and exerts its mechanism of action. Atlas of Science 2016; http://www.atlasofscience.org/

111) Sobell HM. Might trace amounts of actinomycin given over an extended period of time prove to be a powerful anticancer chemotherapeutic regimen? *World J Pharm Res* 2017; **6**: 78-84.

112) Sobell HM. The Centers of Premeltons Signal the Beginning and Ends of Genes. *Biochem and Mol Biol J.* 2017; 3: 17.

113) Sobell H.M. Might Actinomycin be Used to Cure All Cancer? *Cancer Stud Ther J.* 2019; 4(1) 1-3